BIRD STORIES

鸟的故事

[美] 伊迪斯 · M. 派奇 著　亢海宏　李鲁闽 译

天津出版传媒集团
百花文艺出版社

图书在版编目（CIP）数据

鸟的故事 /（美）伊迪斯·M.派奇著；亢海宏，李鲁闽译.
-- 天津：百花文艺出版社，2016.7(2022.1 重印)
ISBN 978-7-5306-6915-0

Ⅰ.①鸟… Ⅱ.①伊… ②亢… ③李… Ⅲ.①鸟类-儿童读物 Ⅳ.①Q959.7-49

中国版本图书馆 CIP 数据核字(2016)第 166799 号

鸟的故事
NIAO DE GUSHI
(美) 伊迪斯·M.派奇 著

出 版 人：薛印胜
选题策划：杨进刚　**美术编辑**：郭亚红
责任编辑：李　跃　**封面设计**：明轩文化 TEL:23674746 · 王　烨
出版发行：百花文艺出版社（天津）有限公司
地址：天津市和平区西康路 35 号　**邮编**：300051
电话传真：+86-22-23332651（发行部）
+86-22-23332656（总编室）
+86-22-23332478（邮购部）
网址：http://www.baihuawenyi.com
印刷：三河市同力彩印有限公司
开本：880×1230 毫米　1/32
字数：254 千字
印张：13.125
版次：2016 年7 月第1 版
印次：2022 年1 月第3 次印刷
定价：49.80元

如有印装质量问题，请与三河市同力彩印有限公司联系调换
地址：河北省三河市黄庄镇赵家庄同力彩印
电话：17710717619

目录

1

Chick, D.D.
山雀奇客 D.D.:圣诞树王国里的小牧师

Right in the very heart of Christmas-tree Land there was a forest of firs that pointed to the sky as straight as steeples. A hush lay over the forest, as if there were something very wonderful there, that might be meant for you if you were quiet and waited for it to come. Perhaps you have felt like that when you walked down the aisle of a church, with the sun shining through the lovely glass in the windows. Men have often called the woods "temples"; so there is, after all, nothing so very strange in having a preacher live in the midst of the fir forest that grew in Christmas-tree Land.

在“圣诞树王国”的最中心有一片枞树,树尖笔直挺拔,像教堂的塔尖,直指天空。枞树林里悄无声息,像藏着什么神奇的东西。嘘!别出声,静静地等着吧,那东西可能就是给你的。也许当你走在教堂的甬道上,阳光透过色彩斑斓的玻璃窗洒满你的全身时,你曾有过这样的感觉。人们常常把树林称作“神殿”;所以,“圣诞树王国”的这片枞树林里住着一位牧师也就无须大惊小怪了。

And the sermon itself was not very strange, for it was about peace and good-will and love and helping the world and being happy—all very proper things to hear about while the bells in the city churches, way, way off, were ringing their glad messages from the steeples.

要说这位牧师的布道,其实也没什么好奇怪的,讲的无非都是和平、亲善和友爱,还有如何帮助这个世界和做一个幸福的人——城市教堂的塔尖上,钟声远远地响起,送来一阵阵福音,这个时候聆听这些布道再合适不过了。

But the minister was a queer one, and his very

first words would have made you smile. Not that you would have laughed at him, you know. You would have smiled just because he had a way of making you feel happy from the minute he began.

不过，这位牧师很特别，他一张口，就会惹你笑；喂，我可不是说你会嘲笑他，我是说，从他张口的那一刻起，不知为什么他就会让你感到很开心，你就会不由自主地笑起来。

He sat on a small branch, and looked down from his pulpit with a dear nod of his little head, which would have made you want to cuddle him in the hollow of your two hands.

他坐在一根小小的树枝上，低头注视着讲坛下的一切，点一下小脑袋，神情可爱极了，让你忍不住想伸出双手捧住他。

His robe was of gray and white and buff-colored feathers, and he wore a black-feather cap and bib.

他身着灰、白和暗黄三色长羽袍，戴着一顶黑色的羽帽，披着一条黑色的羽毛挂肩。

Firs that pointed to the sky.
直指云天的枞树林。

He began by singing his name. "Chick, D.D.," he called. Now, when a person has "D.D." written after his name, we have a right to think that he is trying to live so wisely that he can teach us how to be happier, too. Of course Minister Chick had not earned those letters by studying in college, like most parsons; but he had learned the secret of a happy heart in his school in the woods.

他的布道从自报姓名开始。“奇客 D.D.”,他大声唱道。喂,如果一个人在他的名字后面加上“D.D.”(神学博士),那我们就能做如下推测:他正在努力打造睿智人生,所以,他可以告诉我们怎样才能更快乐。当然了,和多数牧师不同,奇客牧师的这些学识可不是在大学里得来的;他拥有快乐之心的秘方是在他的学校——树林里学到的。

Yes, he began his service by singing his name; but the real sermon he preached by the deeds he did and the life he lived. So, while we listen to his happy song, we can watch his busy hours, until we are acquainted with the little black-capped minister who called himself "Chick, D.D.".

是的,他的布道从自报姓名开始。可是,他真正的布道却是通过行动和生活方式来进行宣讲的。所以,我们在聆听他快乐的鸣唱时,还可以观察他那忙忙碌碌的日子,直到和这位自称"奇客 D.D."的黑顶小牧师相识相知。

Chick´s Christmas–trees were decorated, and no house in the whole world had one lovelier that morning than the hundreds that were all about him as far as he could see. The dark–green branches of the pines and cedars had held themselves out like arms waiting to be filled, and the snow had been dropped on them in fluffy masses, by a quiet, windless storm. It had been very soft and lovely that way—a world all white and green below, with a sky of wonderful blue that the firs pointed to like steeples. Then, as if that were not decoration enough, another storm had come, and had put on the glitter that was brightest at the edge of the forest where the sun shone on it. The second storm had covered the soft white with dazzling ice. It had swept across the white–barked birch trees and their purple–brown branches, and had left them shining all over. It had dripped icicles from the tips of all the twigs that now shone in the sunlight brighter than candles, and tinkled like little bells, when the breezes clicked them together, in a tune that is called, "Woodland Music after an Ice-Storm."

奇客的圣诞树都已经装饰一新，在他眼里，那天早晨整个世界所有房子里的圣诞树都没有他周围那数以百计的圣诞树漂亮。松树和西洋杉伸出墨绿的树枝，像一只只期待着礼物的手臂。一场大雪悄无声息地将雪花撒落在这些伸出的树枝上，为它们盖上了一团团茸茸的积雪，那景象看上去可真是又温馨又漂亮：下面是白茫茫、绿茵茵的双色世界，上面是碧蓝、碧蓝的天空，枞树则像塔尖一样直指着蓝天。然而，这样的装饰似乎还不够美，于是，又下了一场大雪，把阳光照耀下的森林边缘装点成了最晶莹、最亮眼的一道风景。这第二场大雪为松软的积雪覆上了一层耀眼炫目的冰。它还漫过白桦树那包裹着白色树皮的树干，漫过那棕褐色的树枝，让它们全身上下都开始闪闪发光。它让所有的嫩枝尖都滴下了冰柱，现在那些冰柱正在阳光下熠熠闪耀，比烛光还要明亮；微风吹来，冰柱撞在一起，像小铃铛一样叮当作响，就像是在奏乐，乐曲的名字就叫“冰暴过后的森林之乐”。

That is the tune that played all about the black-

Woodland Music after an Ice-Storm.
冰暴过后的森林之乐。

capped bird as he flitted out of the forest, singing, "Chick, D.D.," as he came. The clear cold air and the exercise of flying after his night´s sleep had given Chick a good healthy appetite, and he had come out for his breakfast.

当这位黑顶的鸟儿像来的时候那样唱着"奇客, D. D.",掠过树丛,飞出森林的时候,他的四周到处都在演奏这首乐曲。昨天睡了一个好觉,今天森林里空气又清新凛冽,再加上不时地飞来飞去,让奇客胃口大开,这不,他出来找早餐吃了。

He liked eggs very well, and there were, as he knew, plenty of them on the birch trees, for many a time he had breakfasted there. Eggs with shiny black shells, not so big as the head of a pin; so wee, indeed, that it took a hundred of them or more to make a meal for even little Chick.

奇客特别喜欢吃卵,他还知道白桦树上就有许多卵,因为他已经在那里吃了很多次早餐了。这些卵的外壳亮亮的、黑黑的,还没有大头针的针头大;的确,它们太小了,甚至连小小的奇客一顿都要吃上一百多个。

But he wasn´t lazy. He didn´t have to have eggs cooked and brought to his table. He loved to hunt for them, and they were never too cold for him to relish; so out he came to the birch trees, with a cheery "Chick, D.D.", as if he were saying grace for the good food tucked here and there along the branches.

不过,奇客并不懒惰。他无须找人把卵做熟端到他的餐桌上。他喜欢自己去寻卵吃,而且,尽管这些卵吃起来凉冰冰的,却从来也不影响他的食欲。所以,他出来,飞进了白桦林,兴高采烈地唱着"奇客 D.D.",好像在感谢他能吃到藏在树枝各处的美食。

When he alighted, though, it wasn´t the bark he found, but a hard, thick coating of ice. The branches rattled together as he moved among them and the icicles that dangled down rang and clicked as they struck one another. The ice-storm had locked in Chick´s breakfast eggs, and, try as he would with his little beak, he couldn´t get through to find them.

可是,奇客落下来站定时,才发现他的脚下不是树皮,而是一层硬邦邦、厚墩墩的冰。他在树枝间穿梭跳跃的时候,树枝就吱吱咯咯响作一团,悬挂着的冰柱相互撞击,也叮当作响。冰暴已经把奇客的早饭锁起来了,无论他的小嘴巴怎么使劲地凿，都无法凿穿冰层找到那些卵。

So Chick´s Christmas Day began with hardship: for, though he sang gayly through the coldest weather, he needed food to keep him strong and warm. He was not foolish enough to spend his morning searching through the icy birch trees, for he had a wise little brain in his head and soon found out that it was no use to stay there. But he didn´t go back to the forest and mope about it. Oh, no. Off he flew, down the short hill slope, seeking here and there as he went.

就这样，奇客的圣诞节刚开始就遇到了麻烦：虽然，天寒地冻的日子里，他一直在欢快地鸣唱，可是他也需要食物保持体力，还有体温。奇客不会笨得整个早晨都在冰天雪地的白桦林里觅食，他的小脑瓜聪明着呢，很快他就发现继续待在这里毫无用处。不过，奇客并没有返回森林，也没有因此而闷闷不乐。哦，不会的。他飞走了，沿着小山坡往下飞去，一边飞还一边四处寻觅。

Where the soil was rocky under the snow, some

sumachs grew, and their branches of red berries looked like gay Christmas decorations. The snow that had settled heavily on them had partly melted, and the soaked berries had stained it so that it looked like delicious pink ice-cream. Some of the stain had dripped to the snow below, so there were places that looked like pink ice-cream there, too. Then the ice-storm had crusted it over, and now it was a beautiful bit of bright color in the midst of the white-and-green-and-blue Christmas.

积雪下岩石比较多的地方，长着一些漆树，一枝枝红红的漆果看起来就像是欢乐喜庆的圣诞装饰。树枝上厚厚的积雪已经开始融化，浸透雪水的漆果染红了积雪，让这些正在融化的积雪看起来就像是美味的粉色冰激凌。一些染红的雪水滴落到下面的积雪上，所以，那儿有的地方也变成了粉色冰激凌。接着，冰暴又把这些粉色冰激凌变成了硬硬的冰壳，为这个白茫茫、绿茵茵、蓝莹莹的圣诞大世界点缀出一抹亮丽。

Chick stopped hopefully at the sumach bushes, not because he knew anything about ice-cream or cared a great deal about the berries; but sometimes there were plump little morsels hidden among them, that he liked to pull out and eat. If there was anything there that morning, though, it was locked in under the ice; and Chick flew on to the willows that showed where the brook ran in summer.

奇客满怀希望地落在漆树丛中，这倒不是因为他知道这里有冰激凌，也不是他对漆果特别感兴趣，而是那儿有时会藏着一些肥嘟嘟的小零食，他喜欢拽出来吃上一口。

不过这天早晨即使有吃的，也是被锁在了冰壳里的；于是，奇客又继续飞，来到了柳树林里。有柳树的地方，夏天肯定会有潺潺的小溪流过。

Ah, the willow cones! Surely they would not fail him! He would put his bill in at the tip and down the very middle, and find a good tasty bit to start with, and then he would feel about in other parts of the cone for small insects, which often creep into such places for the winter. The flight to the willows was full of courage. Surely there would be a breakfast there for a hungry Chick!

啊！柳树果！它们肯定不会让他失望的！他要把嘴巴插进果尖，一直插到球果的中心，先在那里找到一点儿好吃的，然后，再去球果的其他部位摸索搜寻，因为那种地方常常有爬进来过冬的小昆虫。飞到柳树林可是需要勇气的，所以，那儿肯定会有一顿早餐，在等着饥肠辘辘的奇客享用吧！

But the ice was so heavy on the willows that it had bent them down till the tips lay frozen into the crust below.

然而，柳树上的冰太厚了，把树都压弯了，连柳梢都躺在树下的冰壳里冻僵了。

So from pantry to pantry Chick flew that morning, and every single one of them had been locked tight with an icy key. The day was very cold. Soon after the ice-storm, the mercury in the thermometer over at the Farm-House had dropped way down below the zero mark, and the wind was in the north. But the cold did not matter if Chick could find food. His feet were bare; but that did not matter, either, if he could eat. Nothing mattered to the brave little black-capped fellow, except that he was hungry, oh, so hungry! And he had heard no call from anywhere to tell him that any other bird had found a breakfast, either.

所以，这天早晨，奇客在各个食品储藏间辗转寻觅，可是，所有的储藏间上都挂着一把冰凉刺骨的锁。这天冷极了。冰暴过后不久，那边“农舍”里温度计的水银柱就一个劲儿地下降，已经降到零度以下了，而且，风也是从北边刮来的。不过，要是奇客能找到食物，这点寒冷就不算什么；他还光着脚丫子，不过如果有饭吃，这也不算什么。对这个黑顶的小勇士来说什么都不在话下，除了一样：饿肚子。唉，太饿了！不过，他也没有接到其他地方的消息，说有哪只鸟找到了一顿早餐。

No, the birds were all quiet, and the distant church-bells had stopped their chimes, and the world was still. Still, except for the click of the icicles on the twigs when Chick or the wind shook them.

没有，所有的鸟儿都噤声了，远处教堂的钟声也渐渐消散，整个世界都悄无声息。偶尔只有奇客和风儿晃动树枝时，冰柱发出的咔嗒声。

Then, suddenly, there was a sound so big and deep that it seemed to fill all the space from the white earth below to the blue sky above. A roaring

BOOOOOOOM, which was something like the waves rushing against a rocky shore, and something like distant thunder, and something like the noise of a great tree crashing to the earth after it has been cut, and something like the sound that comes before an earthquake.

突然，一声巨响腾空而起，又大又沉，下到白茫茫的大地，上到蓝莹莹的天空，每一丝空间都在震颤。那声音“隆隆隆……”地咆哮而来，如巨浪拍打着岩石嶙峋的海岸，如远处滚滚的惊雷，如被砍断的大树轰然倒地时的巨响，如地震前的轰鸣。

It is not strange that Chick did not know that sound. No one ever hears anything just like it, unless he is out where the snow is very light and very deep and covered with a crust.

奇客不知道这是什么声音并不奇怪，谁也没有听过这样的声音，除非他待在外面的那个地方，那里地上的雪很轻很轻，也很厚很厚，上面还结着一层硬硬的冰壳。

Then, if the crust is broken suddenly in one place, it may settle like the top of a puffed-up pie that is pricked; and the air that has been prisoned under the crust is pushed out with a strange and mighty sound.

要知道,如果某个地方的冰壳突然破了,情形大概就像一个胀满气的馅饼被戳破了一样,困在冰壳里面的空气会被挤出来,发出奇怪的巨响。

So that big BOOOOOOOM meant that something had broken the icy crust which, a moment before, had lain over the soft snow, all whole, for a mile one way and a mile another way, and half a mile to the Farm-House.

所以,这"隆隆隆……"的巨响意味着有什么东西把冰壳给弄破了,那冰壳刚才还好好地铺在又松又软的积雪上呢,现在却整个都破了,这边一英里,那边一英里,到了农舍那边又是半英里。

Yes, there was the Farmer Boy coming across the field, to the orchard that stood on the sandy hillside near the fir forest. He was walking on snowshoes, which cracked the crust now and then; and twice on the way to the orchard he heard a deep BOOOOOOOM, which he loved just as much as he loved the silence of the field when he stopped to listen now and then. For the winter sounds were so dear to the Farmer Boy who lived at the edge of Christmas-tree Land, that he would never forget them even when he should become a man. He would always remember the snowshoe tramps across the meadow; and in after years, when his shoulders held burdens he could not see, he would remember the bulky load he carried that morning without minding the weight a bit; for it was a big bag full of Christmas gifts, and the more heavily it pressed against his shoulder, the lighter his heart felt.

的确如此。农家小男孩正穿过田野，往果园那边走。果园就在半山腰的一块沙地上，紧挨着杉树林。男孩穿着雪地鞋行走在冰壳上，时不时地将冰壳踩裂；有两次，在去往果园的路上他听见了低沉的隆隆声，他时不时停下脚步仔细倾听，他喜欢极了这种声音，就像他也喜欢田野的寂静一样。对于住在“圣诞树王国”边上的农家男孩来说，这些冬之音实在是太美妙了，即使他将来长大成人，变成顶天立地的男子汉时也不会忘记这声音。雪地鞋踏过草地发出的咯吱声将永远留在他的记忆里；今后的人生岁月里，当他肩负无法预料的重担时，他会想起那天早晨他身背重重的行李，却毫不在意；因为那是一只装满了圣诞礼物的大口袋，所以，压在肩上的口袋越重，他的心情就越轻快。

When he reached the orchard, he dropped the bag on the snow and opened it. Part of the gifts he spilled in a heap near the foot of a tree, and the rest he tied here and there to the branches. Then he stood still and whistled a clear sweet note that sounded like "Fee–bee".

来到果园后，男孩把口袋放到雪地上打开。他把一部分礼物倒到一棵树下堆成了一堆，把剩下的那部分礼物这儿一个，那儿一包地系在了树枝上。然后，他站好，吹了一声口哨，那声音清脆悦耳，就像是“菲——比”。

Now, Chick, over by the willows had not known what BOOOOOOOM meant, for that was not in his language. But he understood "Fee–bee" in a minute, although it was not nearly so loud. For those were words he often used himself. They meant, perhaps, many things; but always something pleasant. "Fee–bee" was a call he recognized as surely as one boy recognizes the signal whistle of his chum.

奇客在柳树林上辗转奔波时，不知道那“隆隆……”的巨响是什么意思，因为他的语言里没有这种声音。可是，尽管“菲——比”这一声口哨算不上是巨响，他却一下子就明白了。因为这句话是他自己经常说的！这句话也许有很多意思，但都与好事相关。“菲——比”是在呼唤他，他一下子就听出来了，非常肯定，肯定得就如同男孩们能毫不费力地听出同伴用口哨发出的信号一样。

So, of course, Chick flew to the orchard as quickly as he could and found his present tied fast to a branch. The smell of it, the feel of it, the taste of it, set him wild with joy. He picked at it with his head up, and sang "Chick, " He picked at it with his head down and called, "Chick, Chick, D.D." He flew here and there, too gay with happiness to stay long anywhere, and found presents tied to other branches, too. At each one he sang "Chick, Chick, Deee Deeee." It was, "indeed" the song of a hungry bird who had found good rich suet to nibble.

所以，奇客拼尽全力飞到了果园里，看到给自己的礼物紧紧地系在了一根树枝上。哦，它的气息，它的感觉，它的味道，简直让他欣喜若狂！他抬头吃一口唱一句“奇客”，又低头吃一口叫一句“奇客，奇客 D.D.”。他太高兴啦，飞来飞去，哪里都无法久待，结果发现别的树枝上也系着礼物。每发现一个礼物，奇客都会乐得大唱：“奇客，奇客 D.D.，奇客，奇客 D——D——”这的确是饥肠辘辘的小鸟发现有美味、营养的板油可以细嚼慢咽的时候发自内心的欢歌！

The Farmer Boy smiled when he heard it, and waited, for he thought others would hear it, too. And they did. Two birds with black–feather cap and bib heard it and came; and before they had had time to go frantic with delight and song, three others just like them came, and then eight more, and by that time there was such a "Chick"–ing and "D.D."–ing and such a whisking to and fro of black caps and black bibs, that no one paid much attention when Minister Chick, D. D., himself, perched on a branch for a minute, and gave the sweetest little warble that was ever heard on a winter´s day. Then he whistled "Fee–bee" very clearly, and went to eating again, heeding the Farmer Boy no more than if he were not there at all.

听到这首欢歌，农家小男孩笑了；他还在等待，因为他觉得其他的鸟儿也会听到这欢乐的鸣唱。他们真的听到了。两只戴着黑色羽帽，披着黑色挂肩的鸟儿听到歌声飞过来了；他们还没来得及欣喜若狂、高声欢歌的时候，另外三只一模一样的鸟儿也飞来了，接着又来了八只，不一会儿，到处是都是“奇客”、“D——D——”之声，到处是黑冠、黑挂肩上下翻飞的身影。已经没人注意奇客D.D.牧师了，他在一根树枝上待了一会儿，唱了一首歌，这首歌婉转悠扬，是冬天所能听到的最优美的歌了。接着，他吹了一声清脆的口哨“菲——比”，然后，就又开始埋头大吃起来，全然忘了那个农家男孩，好像他根本就不存在似的。

And he wasn´t there very long; for he was hungry, too; and that made him think about the good whiff he had smelled when he went through the kitchen with the snowshoes under his arm, just before he strapped them over his moccasins outside the door.

男孩也没待多久，因为他的肚子也咕咕叫了。这让他想起了刚才从厨房经过时闻到的那股香喷喷的味道，当时他正夹着雪地鞋打算到屋外去把它们扎在鹿皮软鞋上呢。

Yes, that was the Farmer Boy going away with a clatter over the snow-crust; but who were these coming through the air, with jerky flight, and with a jerky note something like "Twitterty-twit-twitterty-twit-twitterty-twitterty-twitterty-twit"? They flew like goldfinches, and they sounded like goldfinches, both in the twitterty song of their flight and their "Tweeet" as they called one another. But they were not goldfinches. Oh, my, no! For they were dressed in gray, with darker gray stripes at their sides; and when they scrambled twittering down low enough to show their heads in the sunlight, they could be seen to be wearing the loveliest of crimson caps, and some of them had rosy breasts.

没错,那是农家小男孩离开时踩在冰壳上发出的咔嗒声;可是,从空中正往这里赶的又是谁呢?他们一边急急忙忙地飞着,一边还急急忙忙地说着什么“推特推——推特——推特推——推特——推特推——推特——”飞起来样子像是金翅雀,叫声也像金翅雀,因为,飞的时候他们唱“推特推”,彼此呼唤的时候他们唱“推——特”哎呀,天哪!不是金翅雀!因为他们穿着灰色的衣服,两侧还有深灰的条纹;当他们叽叽喳喳争先恐后落到低处,脑袋上洒满阳光时,就可以看清楚了,他们戴着深红色的头冠,是最漂亮的那种,有些鸟儿的胸脯还是玫瑰色的。

The redpolls had come! And they found on top of the snow a pile of dusty sweepings from the hay-mow, with grass-seeds in it and some cracked corn and crumbs. And there were squash-seeds, and sunflower-seeds, and seedy apple-cores that had been broken up in the grinder used to crunch bones for the chickens; and there were prune-pits that had been cracked with a hammer.

红顶雀来了!他们发现了雪地上的那堆灰乎乎的东西，那堆东西是从干草堆里扫出来的，里面有草籽,有一些碎玉米和面包屑,有南瓜籽、向日葵籽和碾子碾碎的粗粒苹果核，这碾子是用来给小鸡碾碎骨头吃的,那堆东西里还有用锤子敲碎的李子核。

The joy-songs of the birds over the suet and seeds seemed a signal through the countryside; and before long others came, too.

鸟儿们看到板油和种子时发出的欢叫,像是一个信号,传遍了四野八乡;很快,其他鸟儿也都纷纷赶过来了。

Among them there was a black-and-white one, with a patch of scarlet on the back of his head, who called, "Ping," as if he were speaking through his nose. There was one with slender bill and bobbed-off tail, black cap and white breast, grunting, "Yank yank," softly, as he ate.

其中有一只黑白两色的鸟儿，后脑勺有一块地方鲜红鲜红的，叫起来“砰”的一声，好像是从鼻孔里哼出来的。还有一只鸟儿，嘴巴细长细长，尾巴秃秃的，头顶黑黑的，胸脯白白的，一边吃，一边还低声咕哝着“杨克——杨克”。

But there was none to come who was braver or happier than Chick, D.D., and none who sang so gayly. After that good Christmas feast he and his flock returned each day; and when, in due time, the ice melted from the branches, it wasn´t just suet they ate. It was other things, too.

不过，所有来这里的鸟儿中，谁也没有奇客勇敢、没有奇客开心，谁都没有奇客唱得这么欢快。吃完那顿圣诞节大餐后，奇客和伙伴们每天都会重新返回果园；又过了一段时间，树枝上的冰雪开始消融，他们吃的可就不只这些板油了，还有其他东西呢。

That is how it happened that when, early in the spring, the Farmer Boy examined the apple-twigs, to see whether he should put on a nicotine spray for the aphids and an arsenical spray for the tent caterpillars, he couldn´t find enough aphids to spray or enough caterpillars, either. Chick, D.D. and his flock had eaten their eggs.

于是就发生了这样的事情：早春，农家小男孩来检查苹果树的嫩枝，看看需不需要喷洒尼古丁溶液消灭蚜虫，或者砒霜溶液对付天幕毛毛虫时，他发现需要消灭的蚜虫和需要对付的毛毛虫并不多，奇客 D.D.和他的伙伴们早就把这些害虫的卵吃掉了。

Again, late in the summer, when it was time for the yellow-necked caterpillars, the red-humped caterpillars, the tiger caterpillars, and the rest of the hungry crew, to strip the leaves from the orchard, the Farmer Boy walked among the rows, to see how much poison he would need to buy for the August spray. And again he found that he needn´t buy a single pound. Chick, D.D. and his family were tending his orchard!

同样，夏末，正是黄颈毛虫、红疣毛虫、虎斑毛虫和其他害虫饥饿难耐开始大肆啃噬果园里的叶子的时候，农家小男孩在一排排果树林之间行走查看，想确认一下八月份需要喷洒多少农药。结果，他再一次发现，他连一磅都不用买。山雀奇客 D.D.和他的家族一直在照看他的果园呢！

Yes, Minister Chick was a servant in the good world he lived in. He saved leaves for the trees, he saved rosy apples for city girls and boys to eat, and he saved many dollars in time and spray–money for the Farmer Boy.

是的，奇客牧师一直在为这个他生活其间的美好世界尽心尽力。他拯救了树上的叶子，他让城里的男孩和女孩们吃到了红彤彤的苹果，他让农家男孩节省了很多喷洒农药的时间和金钱。

And all he charged was a living wage: enough suet in winter to tide him over the icy spells, and free house–rent in the old hollow post the Farmer Boy had nailed to the trunk of one of the apple trees.

他做这一切都是分文不取，只要能维持基本生活就行：冬天有足够的板油让他熬过那段冰天雪地的日子，还能免费居住在农家男孩钉在一棵苹果树干上的空心老木桩里。

That old hollow post was a wonderful home. Chick, D.D. had crept into it for the first time Christmas afternoon, when he had eaten until dusk overtook him before he had time to fly back to the shelter of the fir forest. He found that he liked that post. Its walls were thick and they kept out the wind; and, besides, was it not handy by the suet?

那个空心老木桩住着还真不错。圣诞节那天下午，他一直在吃呀吃的，等夜幕降临的时候，他已经没有时间返回位于杉树林的住所了，于是，他第一次钻了进去。结果，他发现自己很喜欢这个空心老木桩：厚厚的四壁，抵挡了风寒；另外，木桩跟前不就是板油吗？

In the spring he liked it for another reason, too—the best reason in the world. It gave great happiness to Mrs. Chick. "Fee–bee?" he had asked her as he called her attention to it; and "Fee–bee," she had replied on looking it over. So he said, "Chick, D.D." in delight, and then perched near by, while he warbled cosily a brief song jumbled full of joy.

春天，他喜欢空心老木桩还另有因由，这是全世界最充分的因由：这个木桩让奇客太太快乐无比。奇客把木桩指给太太看时，征询道："菲——比？"；太太只看了一眼，就回答说："菲——比。"于是，奇客兴奋地说了声"奇客，D.D."，就飞过去落在木桩旁边，一边飞还一边悠然自得、兴高采烈地一通乱唱。

Chick and his mate had indeed chosen well, for it is a poor wall that will not work both ways. If the sides of the hollow post had been thick enough to keep out the coldest of the winter cold, they were also thick enough to keep out the hottest of the summer heat. If they kept out the wet of the driving storm, they held enough of the old-wood moisture within so that the room did not get too dry. Of course, it needed a little repair. But then, what greater fun than putting improvements into a home? Especially when it can be done by the family, without expense!

奇客和他的太太真的很会选，因为只有两头兼顾的墙壁才算是好的墙壁。如果空心木桩厚得都抵御住了最凛冽的严冬，那么也一定可以抵挡最炎热的酷暑；如果他们能把瓢泼大雨拦在门外，那么也一定能把老木头的湿润留在屋内，让屋子不会太干燥。当然啦，需要小小地修缮一下。可是，还有什么比修缮自己的小窝更有趣的呢？尤其是全家一起上阵，还不花费一分一毫！

So Mr. and Mrs. Chick fell to work right cheerily, and dug the hole deeper with their beaks. They didn´t leave the chips on the ground before their doorway, either. They took them off to some distance, and had no heap near by, as a sign to say, "A bird lives here." For, sociable as they were all winter, they wanted quiet and seclusion within the walls of their own home.

于是，奇客夫妇立刻欢天喜地地行动起来。他俩用嘴巴把洞啄得更深一些，可没有把木屑丢在门前的地上，而是把他们衔到稍远的地方，免得在家门附近堆成小堆儿，那就等于在说："一只鸟儿住在这里。"尽管整个冬天他俩都乐于交际，但是，他们也需要关起门来享受家的静谧和独处的乐趣。

And such a home it was! After it had been hollowed to a suitable depth, Chick had brought in a tuft of white hair that a rabbit had left among the brambles.

Mrs. Chick had found some last year´s thistle–down and some this year´s poplar cotton, and a horse–hair from the lane. Then Chick had picked up a gay feather that had floated down from a scarlet bird that sang in the tree–tops, and tore off silk from a cocoon. So, bit by bit, they gathered their treasures, until many a woodland and meadow creature and plant had had a share in the softness of a nest worthy of eight dear white eggs with reddish–brown spots upon them. It was such a soft nest, in fact, with such dear eggs in it, that Chick brooded there cosily himself part of the time, and was happy to bring food to his mate when she took her turn.

这个家可真别致啊！等把洞掏得深浅合适的时候，奇客就从荆棘丛里带回来一簇白兔毛。奇客太太从下面找回一些去年的蓟草和今年的杨絮，还从路边捡来一根马尾毛。后来，一只猩红色的小鸟在树梢唱歌的时候抖落了一根羽毛，很是鲜艳，奇客也捡了回来。他还从蚕茧上扯回了几根蚕丝。就这样，他俩一点、一点地往家里搜罗奇珍异宝，他们的窝大功告成了，可以舒舒服服地安顿八颗白地红棕斑点的宝贝鸟蛋了，这片林地和草场上的许多动物、植物都或多或少地为这个柔软的窝尽了一份力。这个小窝这么松软，里面的鸟蛋这么珍贵，让奇客有一半的时间留在家里优哉游哉地自己孵蛋；即使轮到太太孵蛋，奇客也会满心欢喜地为她觅食送饭。

In eleven or twelve days from the time the eggs were laid, there were ten birds in that home instead of two. The fortnight that followed was too busy for song. Chick and his mate looked the orchard over even more thoroughly than the Farmer Boy did; and before those eight hungry babies of theirs were ready to leave the nest, it began to seem as if Chick had eaten too many insect eggs in the spring, there were so few caterpillars hatching out. But the fewer there were, the harder they hunted; and the harder they hunted, the scarcer became the caterpillars. So when Dee, Chee, Fee, Wee, Lee, Bee, Mee, and Zee were two weeks old, and came out of the hollow post to seek their own living, the whole family had to take to the birches until a new crop of insect eggs had been laid in the orchard. This was no hardship. It only added the zest of travel and adventure to the pleasure of the days. Besides, it isn´t just orchards that Chick, D.D. and his kind take care of. It is forests and shade–trees, too.

产下鸟蛋的第十一天或第十二天，小窝里就不是两只鸟儿，而是十只了。接下来的两周他们忙得连歌都没时间唱了。奇客和太太在果园里到处检查，比农家男孩检查得还彻底；春天，他们的这八只嗷嗷待哺的小宝宝还没准备好离家自食其力的时候，奇客好像是吃掉了太多的昆虫卵，结果孵出来的毛毛虫寥寥无几。可是，毛毛虫越少，他们就越努力地去寻觅；他们越努力地去寻觅，毛毛虫就越少。所以，当刚出生两个星期的嘀嘀、叽叽、菲菲、喂喂、哩哩、吡吡、咪咪和啧啧从空心木桩里出来独闯天下的时候，他们不得不举家迁往桦树林，直到果园里生出新一茬的昆虫卵。举家迁移没什么可辛苦的，只不过是在白天的快乐中又增添了些旅行的兴奋和冒险的刺激罢了。再说了，奇客 D.D.和伙伴们悉心呵护的不光是果园，还有森林和那些遮荫树呢。

Hither and yon they hopped and flitted, picking the weevils out of the dead tips of the growing pine trees, serving the beech trees such a good turn that the beechnut crop was the heavier for their visit, doing a bit for the maple–sugar trees, and so on through the woodland.

他们在林地里蹦来蹦去，蹿上蹿下，好事做了一件又一件：把小松树枯梢里的象鼻虫捉出来；精心照料山毛榉，结果因为他们的造访，山毛榉今年结的坚果最多；为枫糖树出诊等等，真是不胜枚举。

Not only did they mount midget guard over the mighty trees, but they acted as pilots to hungry birds less skillful than themselves in finding the best feeding-places. "Chick, D.D.D.D.D.D.", they called in thanksgiving, as they found great plenty; and warblers and kinglets and creepers and many a bird beside knew the sound, and gathered there to share the bountiful feast that Chick, D.D. had discovered.

除了充当大树的小守护神之外，他们还为那些觅食技术不如他们的小鸟充当导航员，把饥肠辘辘的小鸟引到最佳的觅食处。看到那么多的食物，这些鸟儿都十分感激地叫着："奇客，嘀——嘀——"；鸣鸟、戴菊鸟、旋木雀和附近的许多鸟儿都听得懂这种叫声，于是全都聚集而来，分享奇客 D.D.发现的盛宴。

The gorgeous autumn came, the brighter, by the way, for the leaves that Chick had saved. The Bob-o-links, in traveling suits, had already left for the prairies of Brazil and Paraguay, by way of Florida and Jamaica. The strange honk of geese floated down from V-shaped flocks, as if they were calling, "Southward Ho! " The red–winged blackbirds gave a wonderful farewell chorus. Flock by flock and kind by kind, the migrating birds departed.

绚丽多彩的秋天到了，顺便说一句，因为奇客拯救的那些树叶，今年的秋天分外亮丽。穿着旅行服的食米鸟已经上路了，他们要经过佛罗里达和牙买加飞往巴西和巴拉圭的大草原。成群结队排成“V”字形的大雁队伍里传来奇怪的雁鸣，似乎在叫：“嚯！往南飞呀！”红翅黑鹂也齐展歌喉表演了一场精彩的惜别大合唱。就这样，候鸟们各归其类、成群结队地出发了。

WHY?

为什么？

Well, never ask Chick, D.D. The north with its snows is good enough for him. Warblers may go and nuthatches may come. It is all one to Chick. He is not a bird to follow fashions others set.

喂，可千万不要叫奇客D.D.同行。白雪皑皑的北方就是他的天堂。鸣鸟可以离去，五子雀可以到来，可是对于奇客来说，去就是来，来就是去。他可不是一只随波逐流的鸟儿。

This bird-of-the-happy-heart has courage to meet the coldest day with a joyous note of welcome. The winter is cheerier for his song. And, as you have guessed, it is not by word alone that he renders service. The trees of the north are the healthier for his presence. Because of him, the purse of man is fatter, and his larder better stocked. He has done no harm as harm is counted in the world he lives in. It is written in books that, in all the years, not one crime, not even one bad habit, is known of any bird who has called himself "Chick, D.D."

他是一只满心欢喜的鸟儿，他有勇气唱着欢快的旋律去迎接最凛冽的冬日。寒冬也因他的歌声而分外欢快。是的，你猜对啦，他履行职责靠的不光是言语。北方的树因他的出现而分外健康。因为他，人类的钱包变得更鼓，食品储藏室也更满。他从未做过坏事，在他生活的那个世界里，做坏事是要记录在案的。书里明明白白地记载着：多少年来，从来没有任何一个自称"奇客 D. D."的鸟儿犯过罪，哪怕是一次不良记录都没有。

Because the world is always better for his living in it; and because no one can watch the black-capped sprite without catching, for a moment at least, a message of cheer and courage and service, does he not name himself rightly a minister?

因为世界总是因他的存在而更加美好；因为所有审视这位黑顶小精灵的人，每时每刻无不从他那里感受到快乐和勇气的力量，每时每刻无不得到他的帮助，所以，他自称牧师难道不是名副其实吗？

Yes, surely, the little parson who dwells in the heart of Christmas-tree Land has a right to his "D.D.", even though he did not earn it in a college of men.

的确，这个居住在“圣诞树王国”中心的小牧师，虽然他的“神学博士”不是在人类的神学院里获得的，但是这个头衔他当之无愧。

2

The Five Worlds of Larie
海鸥拉瑞的五个世界

Larie was all alone in a little world. He had lived there many days, and had spent the time, minute by minute and hour by hour, doing nothing at all but growing. That one thing he had done well. There is no doubt about that; for he had grown from a one-celled little beginning of life into a creature so big that he filled the whole of his world crammed full. It was smooth, and it was hard, and its sides were curved around and about him so tightly that he could not even stretch his legs. There was no door. Larie was a prisoner. The prison-walls of his world held him so fast that he could not budge. That is, he could not budge anything but his head. He could move that a little.

在一个小小的世界里，躺着孤零零的拉瑞。他已经在那里住了好多天了。时间一分一分、一小时一小时地过去了，他只做了一件事情：发育。这件事情他做得不错，这一点无可置疑，因为他已经从单细胞的萌芽状态发育成了一个全新的生命，把他的小世界塞了个满满当当。他的小世界又滑又硬，四面穹顶，把他裹得紧紧的，连腿都没法伸一下；连门都没有，简直就是囚禁他的牢房。牢房的墙壁紧紧地贴着他，让他除了脑袋之外，无法动弹。也就是说，他只能稍稍动下脑袋。

Now, that is what we might call being in a fairly tight place. But you don´t know Larie if you think he could not get out of it. There are few places so tight that we can´t get out of them if we go about it the right way, and make the best of what power we have. That is just what Larie did. He had power to move his head enough to tap, with his beak, against the wall of his world that had become his prison. So he kept tapping with his beak. On the end of it was a queer little knob. With this he knocked against the hard smooth wall.

这就是我们所说的陷入困境。不过，要是你以为他无法摆脱这个困境，那你就太不了解拉瑞了。只要我们方法得当、竭尽全力，就没有不能摆脱的困境。拉瑞就是这样。他用力扭动脑袋，嘴巴刚好能敲到小牢房的墙上。于是，他开始用嘴巴不停地敲打。他的嘴巴尖上有一个奇怪的小疙瘩，拉瑞敲打又硬又滑的墙壁用的就是它。

"Tap! tip tip! " went Larie´s knob. Then he would rest, for it is not easy work hammering and pounding, all squeezed in so tight. But he kept at it again and again and again. And then at last he cracked his prison–wall; and lo, it was not a very thick wall after all! No thicker than an eggshell!

“嗒！嗒嗒！”拉瑞嘴巴上的小疙瘩开始敲了。敲一下，他就停一会儿，因为地方这么小，这么挤，敲敲打打可不是件容易的事。可他还是坚持不懈，不停地敲呀敲呀，终于，牢房的墙壁被凿穿了！唉，瞧，这墙本来就不怎么厚嘛！还没蛋壳厚呢！

That is the way with many difficulties. They seem

so very hard at first, and so very hopeless, and then end by being only a way to something very, very pleasant.

很多困难都是这样，刚开始的时候似乎艰难无比，希望渺茫，可是，结果却往往令人异常惊喜。

So here was Larie in his second world. Its thin, soft floor and its thick, soft sides were made of fine bright-green grass, which had turned yellowish in drying. It had no roof. The sun shone in at the top. The wind blew over. There had been no sun or wind in his eggshell world. It was comfortable to have them now. They dried his down and made it fluffy. There was plenty of room for its fluffiness. He could stretch his legs, too, and could wiggle his wings against his sides. This felt good. And he could move his head all he cared to. But he did not begin thumping the sides of his new world with it. He tucked it down between two warm little things close by, and went to sleep. The two warm little things were his sister and brother, for Larie was not alone in his nest-world.

就这样，拉瑞来到了他的第二个世界。这里地板薄薄的、软软的，四周的墙厚厚的，也软软的，全都铺着青翠的小草，小草快干了，有点发黄。这里没有屋顶，头顶上是暖洋洋的太阳，还有风儿吹过呢。他的蛋壳世界可没有太阳，也没有风。现在有了它们可真是舒服。它们弄干了他身上的绒毛，把他变得毛茸茸的，现在有的是地方让他浑身都毛茸茸的。也能伸腿了，能扑棱翅膀了。感觉真不错！还可以随心所欲地转动脑袋，不过这次他没有拿脑袋去撞新世界的墙，而是把它埋在身边两个暖暖的小东西之间，睡着了。两个暖暖的小东西就是他的弟弟和妹妹，现在，鸟窝世界里的拉瑞有伴儿了。

The sun went down and the wind blew cold and the rain beat hard from the east; but Larie knew nothing of all this. A roof had settled down over his world while he napped. It was white as sea foam, and soft and dry and, oh, so very cosy, as it spread over him. The roof to Larie´s second world was his mother´s breast.

太阳落山了，风也冷飕飕的，雨从东边噼里啪啦地落了下来。可是，这一切拉瑞却一无所知。因为，就在他打盹的时候，一个屋顶遮住了他的世界。这个屋顶像浪花一样洁白，盖在他的身上软软的、干干的，噢，真舒服。原来，拉瑞第二个世界的屋顶就是他妈妈的胸膛。

The storm and the night passed, and the sun and the fresh spring breeze again came in at the top of the nest. Then something very big stood near and made a shadow, and Larie heard a strange sound. The something very big was his mother, and the strange sound was her first call to breakfast. When Larie heard that, he opened his mouth. But nothing went into it. His brother and sister were being fed. He had never had any food in his mouth in all the days of his life. To be sure, his egg-world was filled with nourishment that he had taken into his body and had used in growing; but he had never done anything with his beak except to knock with the knob at the end of it against the shell when he pipped his way out. What a handy little knob that had been—just right for tapping. But, now that there was no hard wall about him to break, what should he use it for? Well, nothing at all; for the joke of it is, there was no knob there. It had dropped off, and he could never have another.

雨停了，天亮了，太阳又暖洋洋地照到了鸟巢上，和煦的春风也扑面而来。突然，旁边站起一个遮天蔽日的庞然大物，同时，拉瑞听到一个奇怪的声音。原来，那个庞然大物就是他的妈妈，那个奇怪的声音是她的第一声呼唤，她在招呼孩子们吃早餐。拉瑞听到了，立即张开了嘴巴，可是，却什么也没吃到，原来，都喂给了弟弟和妹妹。自从出生以来，他还滴水未进呢。确实，他的蛋壳世界里面装满了营养，他把这些营养都吸收到身体里，利用它们发育成长，可是，嘴巴尖上的那个疙瘩除了帮他敲碎蛋壳、爬到外面之外，还从未做过别的什么事儿呢。多么灵巧的一个小疙瘩啊，正适合敲东西。可是，现在周围没有硬硬的墙可敲了，那还能拿它来干什么呢？呵呵，什么都干不了啦；因为很可笑，嘴巴上的疙瘩没啦，已经掉了，再也长不出第二个了。

Never mind: he could open his beak just as well without it; and by-and-by his mother came again with a second call for breakfast, and that time Larie got his share. After that, there were calls for luncheon and for dinner, and luncheon again between that and supper; and part of the calls were from Mother and part from Father Gull.

不过，不用担心：没有它，拉瑞照样可以张嘴；不久，妈妈又过来喊他们吃早餐了，这次就有拉瑞的份了。早餐之后，又叫他们吃午餐和晚餐，午餐、晚餐和夜宵之间还有加餐，叫他们吃饭的有时候是海鸥妈妈，有时候是海鸥爸爸。

Larie´s second world, it seems, was a place where he and his brother and sister were hungry and were fed. This is a world in which dwell, for a time, all babies, whether they have two legs, like you and Larie, or four, like a pig with a curly tail, or six, like Nata who lived in Shanty Creek. An important world it is, too; for health and strength and growing up, all depend upon it.

拉瑞的第二个世界似乎就是一个和弟弟妹妹一起饿了吃,吃了饿的世界。所有的小宝宝都曾经在这个世界里待过一段时间,不管他们是像你和拉瑞那样的两腿宝宝，还是像卷尾巴猪那样的四腿小宝宝，或者是像“木屋溪”的蜻蜓娜塔那样的六腿小宝宝。这个世界也很重要，因为宝宝一生的健康、体力和发育都取决于它。

There was, however, only a rim of soft fine dry grass to show where Larie´s nest-world left off and his third world began. So it is not surprising that, as soon as their legs were strong enough, Larie and his brother and sister stepped abroad; for what baby does not creep out of his crib as soon as ever he can?

拉瑞的巢窝世界和他的第三个世界之间只隔着一

圈软软的、细细的干草。所以，拉瑞和弟弟妹妹的腿脚刚一有劲儿，就迈出了边界，这一点儿也不奇怪，因为哪个宝宝不是迫不及待地爬下婴儿床的呢？

They could not, for all this show of bravery, feed themselves like the sons of Peter Pan, or swim the waters like Gavia´s two Olairs at Immer Lake. However grown up the three youngsters may have felt when they began to walk, Father and Mother Gull made no mistake about the matter, but fed them breakfasts, dinners, and suppers, and stuffed them so full of luncheons between meals, that the greedy little things just had to grow, so as to be able to swallow all that was brought them.

可是，尽管他们勇气可嘉，却无法像矶鹞彼得·潘的儿子们那样自己养活自己，也无法像潜鸟加菲娅的两个宝宝那样在爱默湖上游泳。三个小家伙儿刚刚才开始蹒跚学步，就觉得自己已经长大成人了，但是，在这件事情上，海鸥爸爸和海鸥妈妈可一点儿也不犯糊涂，他们不停地给孩子们喂食，早餐、晚餐、夜宵，期间还不住地加餐，结果，三个贪吃的小家伙只好拼命地长个儿，这样才能咽得下喂给他们的那些食物。

There were times, certainly, when Larie still felt very much a baby, even though he ran about nimbly enough. For instance, when he made a mistake and asked some gull, that was not his father or mother, for food, and got a rough beating instead of what he begged for!

当然了，尽管拉瑞腿脚已经灵便，可以到处乱跑了，但是，有些时候，他仍然感觉自己像个小宝宝。比如，有一次他弄错了，把别的海鸥当成了爸爸妈妈，向他们讨食吃，结果不仅什么都没吃到，还挨了一顿猛揍！

Oh, then he felt like a forlorn little baby, indeed; for it was not pleasant to be whipped, and that sometimes cruelly, when he didn´t know any better; for all the big gulls looked alike, with their foam-white bodies and their pearl-gray capes, and they were all bringing food; so how could he know who were and who were not his father and mother gull? Well, he must learn to be careful, that was all, and stay where his very own could find and feed him; for gulls can waste no time on the young of other gulls—their own keep them busy enough, the little greedies!

噢，那个时候他感觉自己就像是一个被遗弃的小宝宝，因为挨揍可不是什么让人高兴的事情，而且，有时候，还被揍得特别狠，因为他不吸取教训，总是弄错。没办法，所有的大海鸥长得都差不多：身子像浪花一样洁白，披着珍珠灰的斗篷，还都叼着吃的，所以，他怎么知道谁是自己的鸥爸鸥妈，谁又不是自己的鸥爸鸥妈呢？可是，无论如何，他必须学会小心，没别的办法；还有，他得待在一个自己的亲爸亲妈能找到自己，而且能给自己喂食的地方；因为，海鸥可没工夫给别人家的孩子喂食，自己的孩子已经够他们忙活的了，这些小贪吃鬼们！

Again, Larie must have felt very wee and helpless whenever a big man walked that way, shaking the ground with his heavy step and making a dark shadow as he came. Then, oh, then, Larie was a baby, and hid near a tuft of grass or between two stones, tucking his head out of sight, and keeping quite still as an ostrich does, or,—yes,—as perhaps a shy young human does, who hides his head in the folds of his mother's skirt when a stranger asks him to shake hands.

还有，每当大块头的人类从那边走过来的时候，沉重的脚步声让大地都为之震颤，而且还有一条暗影随身而行，那个时候，拉瑞一定感到了自己的渺小和无助。

那个时候，噢，那个时候拉瑞就是一个小宝宝，把脑袋塞到草丛或石缝里面藏起来，一动不动就跟鸵鸟一样，或者，对了，也许就像是人类害羞的小孩子，陌生人想和他握手，可是他却把脑袋埋在妈妈的裙褶里藏了起来。

But few men trod upon Larie´s island–world, and no man came to do him harm; for *the regulations under the Migratory–Bird Treaty Act prohibit throughout the United States the killing of gulls at any time.* That means that the laws of our country protect the gull, as of course you will understand, though Larie knew nothing about the matter.

不过人类很少踏上拉瑞的海岛世界，更不会有人来伤害他；因为《联邦候鸟法案》(*Migratory–Bird Treaty Act*)规定在美国境内任何时候都不能捕杀海鸥。也就是说，海鸥受国家法律的保护，虽然拉瑞对此一无所知，可是，你当然会明白的，对吧？

Yes, think of it! There was a law, made at Washington in the District of Columbia, which helped take care of little downy Larie way off in the north on a rocky island.

是的，想想吧！毛茸茸的小拉瑞生活在北方很远很远的一个岩岛上，为了帮着照顾他，竟然还制定了一部法律，在哥伦比亚特区华盛顿制定的法律。

I said "helped take care of"; for no law, however good it may be, can more than help make matters right. There has to be, besides, some sort of policeman to stand by the law and see that it is obeyed.

我是说“帮着照顾”；因为无论多好的法律，也只能是起辅助作用。除此之外，必须还要有专门的警察来捍卫这部法律并监督人们遵守这个法律。

So Larie, although he never knew that, either, had a policeman; and the law and the policeman together kept him quite safe from the dangers which not many years ago most threatened the gulls on our coast islands. In those days, before there were gull–laws and gull–policemen, people came to the nests and took their eggs, which are larger than hens´ eggs and good to eat; and people came, too, and killed these birds for their feathers. Then it was that the beautiful stiff wing-feathers, which should have been spread in flight, were worn upon the hats of women; and the soft white breast-feathers, which should have been brooding brownish eggs all spattered over with pretty marks, were stuffed into feather-beds for people to sleep on.

因此，拉瑞有一个警察，不过，他对此也毫不知情。这部法律和这个警察在共同保护他，让他不再遭受几年前沿海岛屿上的海鸥们所面临的那些危险。那个时候，还没有海鸥保护法和海鸥保护警，人们经常拿走海鸥巢里的海鸥蛋，因为海鸥蛋比鸡蛋大，而且很好吃。另外，为了得到海鸥身上的羽毛，人们也来这里捕杀海鸥。于是，那美丽又挺拔的翅羽，本该是用来凭空翱翔的，却被女人们戴在了帽子上；那柔软洁白的胸羽，本该用来孵化洒满漂亮斑点的褐色海鸥蛋的，却被塞进了人类睡觉用的羽绒床。

Well it was for Larie that he lived when he did; for his third world was a wonderful place and it was right that he should enjoy it in safety. When Larie first left his nest and went out to walk, he stepped upon a shelf of reddish rock, and the whole wall from which his shelf stuck out was reddish rock, too. Beyond, the rocks were greenish, and beyond that they were gray. Oh! The reddish and greenish and grayish rocks were beautiful to see when the fog lifted and the sun shone on them.

拉瑞在生活中磨炼着自己,过得健康又快乐。他的第三个世界精彩纷呈,就该让他安然无忧地享受这个精彩的世界。第一次离巢去散步时,他落在了一块突出来的岩石上,这块岩石是淡红色的,与它相连的石壁也是淡红色的。石壁那边是淡绿色的，再往那边是灰白色的。哦！云开雾散,阳光普照,那淡红、淡绿和淡灰色的岩石看上去可真漂亮啊!

But Larie´s island–world was not all rock of different colors: for over there, not too far away to see, was a dark–green spruce tree. Because rough winds had swept over this while it was growing, its branches were scraggly and twisted. They could not grow straight and even, like a tree in a quiet forest. But never think, for all of that, that Larie´s spruce was not good to look upon. There is something splendid about a tree which, though bending to the will of the mighty winds that work their force upon it, grows sturdy and strong in spite of all. Such trees are somehow like boys and girls, who meet hardships with such courage when they are young, that they grow strong and sturdy of spirit, and warm of heart, with the sort of mind that can understand trouble in the world, and so think of ways to help it.

不过,拉瑞的海岛世界不全是色彩斑斓的岩石,瞧,就在那边不远处,有一棵墨绿色的云杉树,因为还没长大就遭到了狂风的摧残,所以它的枝丫已经扭曲变形,寂静的森林里,树可以长得挺拔整齐,可是这里的云杉却不能。尽管如此,你可千万不要觉得拉瑞的云杉不好看。假如一棵树在狂风的怒吼中弯下了腰,可是依然长得顽强茁壮,那么这棵树一定不同寻常。这样的树有点像人类的孩子,如果小时候勇于面对困难,那么长大后其精神就坚不可摧,其心灵就温暖如春,其人也就善解人意,他也就会想方设法去帮助别人。

Yes, perhaps Larie´s tree was an emblem of courage. However that may be, it was a favorite spot on the island. Often it could be seen, that dark, rugged tree, which had battled with winds from its seedling

Birds, too, that had lived in rough winds.
狂风中的小鸟。

days and grown victoriously, with three white gulls resting on its squarish top—birds, too, that had lived in rough winds and had grown strong in their midst.

是的,或许拉瑞的这棵杉树就是勇气的象征。不管是不是,反正大家都喜欢去那里。瞧,这棵饱经风霜的杉树上经常栖息着三只饱经风霜的白色海鸥。杉树迎着怒吼的狂风,从一棵小树苗长成了一棵参天大树,三只海鸥栖息在杉树顶那平坦的树枝上,也迎着怒吼的狂风,茁壮成长。

There was more on the island than rocks and trees. Over much of it lay a carpet of grass. Soft and fine and vivid green it was, of the kind that had been gathered for Larie´s nest and had turned yellowish in drying. Under the carpet, in underground lanes as long as a man´s long arm, lived Larie´s young neighbor-folk—little petrels, sometimes called "Mother Carey´s Chickens."

海岛上不止有岩石和云杉。岛上大部分地方都铺着一层草毯,又细又软,翠绿翠绿的,拉瑞筑巢用的就是这种草,这种草干了以后就变黄了。草毯下面,有很多和人类胳膊一样长的地下甬道,那里住着拉瑞的伙伴——小海燕，有时他们也被叫作“圣母玛丽亚的小鸡”。

There was even more on the island yet: for high on the rocks stood a lighthouse; and the man who kept the signal lights in order was no other than Larie´s policeman himself. A useful life he lived, saving ships of the sea by the power of light, and birds of the sea by the power of law.

岛上的东西还多着呢:高高的岩石上耸立着一座灯塔；控制信号灯的不是别人，正是保护拉瑞的那位警察。他的生活多么有意义啊！他点燃明灯拯救海上的船只;他严格执法拯救海上的鸟儿。

So that was Larie´s third world—an island with a soft rug of bright-green grass, and big shelfy rocks of red and green and gray, and rugged dark-green trees, with white gulls resting on the branches, and a lighthouse with its signal.

这就是拉瑞的第三个世界:海岛。岛上绿草如茵,像铺着一条松软明丽的地毯；巨石嶙峋，红的、绿的、灰的，色彩斑斓；饱经风霜的墨绿色云杉上栖息着白色的海鸥；还有那传递信号的灯塔。

All around and about that island lay Larie´s fourth world—the sea. When his great day for swimming came, he slipped off into the water; and after that it was his, whenever he wished—his to swim or float upon, the wide–away ocean reaching as far as any gull need care to swim or float.

海岛四周是拉瑞的第四个世界：大海。下水游泳的伟大时刻终于来到了，他轻轻一滑就溜进了大海；从此，大海就属于他了，只要他愿意，他就可以尽情地游，尽情地漂。茫茫大海，浩瀚无边，他想游多远就游多远，想漂到哪里就漂到哪里。

All over and above the sea stretched Larie´s fifth world—the air. When his great day for flying came, he rose against the breeze, and his wings took him into that high–away kingdom that lifted as far as any gull need care to fly.

大海上面是拉瑞的第五个世界：天空。展翅翱翔的伟大时刻终于来到了，他迎着和煦的微风离开地面，乘着翅膀飞上了天空。那是一个高无边际的王国，海鸥们可以尽情地飞，想飞多高就飞多高。

Now that Larie could both swim and fly, he was large, and acted in many ways like an old gull; but the feathers of his body were not white, and he did not wear over his back and the top of his spread wings a pearl–gray mantle.

现在，拉瑞既会游，又会飞了，他体型健硕，行为举止俨然像是一只老海鸥；可是，他身上还没有白如浪花的羽毛，后背和飞羽上也没有披珍珠灰的斗篷。

Nor was he given the garb of his father and mother for a traveling suit, that winter when he went south with the others, to a place where the Gulf Stream warmed the water whereon he swam and the air wherein he flew.

那年冬天，和其他海鸥一起南飞的时候，他身上也没有爸爸妈妈那样的旅行服。他们去的那个地方，因为有墨西哥暖流经过，所以，海水暖暖的，天空也暖暖的。

But there came a time when Larie had put off the clothes of his youth and donned the robe of a grown gull. And as he sailed in the breezes of his fifth world, which blew over the cold sea, and across the island with a carpet of green and rocks of red and green and gray,—for he was again in the North,—he was beautiful to behold, the flight of a gull being so wonderful that the heart of him who sees quickens with joy.

可是,这一天终于来了。拉瑞脱去了幼时的衣服,换上了成年海鸥的长袍。他迎着第五个世界的微风轻快地飞着,那微风掠过冰冷的海面,从海岛上那绿草如茵的草毯和红的、绿的、灰的各色岩石上吹过。啊!他又回到了北方!瞧,他飞得多美!无论是谁看了,都会兴奋不已,心跳加速。

Larie was not alone. There were so many with him that, when they flew together in the distance, they looked as thick as snowflakes in the air; and when they screamed together, the din was so great that people who were not used to hearing them put their hands over their ears.

拉瑞并不孤单,和他结伴而行的海鸥多极了,密密麻麻的,从远处看就像是空中飘洒的雪花;他们一起尖叫，声音震耳欲聋，不习惯这声音的人会赶紧用双手捂住耳朵。

And more than that, Larie was not alone; for there sailed near him in the air and floated beside him in the sea another gull, at whom he did not scream, but to whom he talked pleasantly, saying, "me–you," in a musical tone that she understood.

和拉瑞结伴而行的还另有其人呢。无论他在天空翱翔，还是在海里漂游，身边总是陪伴着另一只海鸥，拉瑞从不对她大喊大叫，而总是柔情蜜意地对她说："蜜——友"，听上去就像在唱歌，对此，她心领神会。

Larie and his mate found much to do that spring. One game that never failed to interest them was meeting the ships many, many waves out at sea, and following them far on their way. For on the ships were men who threw away food they could not use, and the gulls gathered in flocks to scramble and fight for this. Children on board the ships laughed merrily to see them, and tossed crackers and biscuits out for the fun of watching the hungry–birds come close, to feed.

Floated beside him in the sea another gull, to whom he talked pleasantly.
另一只海鸥陪伴在他身旁和他一起在海面漂游,和她说话时,他总是柔情蜜意。

那年春天，拉瑞和太太有很多事情要忙。其中有一项他俩永远乐此不疲的运动，那就是越过重重的海浪去迎接海上的船只，然后一路相送到很远的地方。原来，船上的人们会把吃不完的食物扔进海里，于是海鸥们就成群结队、争先恐后地抢食这些食物。看到他们，船上的孩子就会欢呼雀跃，他们把饼干和面包扔向大海，看着这些如饥似渴的鸟儿纷纷飞来觅食，开心极了。

Many a feast, too, the fishermen gave the gulls, when they sorted the contents of their nets and threw aside what they did not want.

还有渔夫赠送的盛宴呢！渔夫清理渔网里的鱼儿时，会把不要的鱼儿扔到一边。

Besides this, Larie and his mate and their comrades picnicked in high glee at certain harbors where garbage was left; for gulls are thrifty folk and do not waste the food of the world.

除此之外，拉瑞夫妇和伙伴们还会兴致勃勃地飞到一些垃圾处理港去野餐，因为海鸥一个个都很会省吃俭用，从不浪费粮食。

From their feeding habits you will know that these beautiful birds are scavengers, eating things which, if left on the sea or shore, would make the water foul and the air impure. Thus it is that Nature gives to a scavenger the duty of service to all living creatures; and the freshness of the ocean and the cleanness of the sands of the shore are in part a gift of the gulls, for which we should thank and protect them.

看他们的觅食习惯，你就知道这些漂亮的鸟儿是食腐动物，专吃海上或岸上那些会让海水变脏、空气变臭的东西。原来，大自然赋予了食腐动物为所有生物服务的职责。从某种程度上讲，清新的海洋和洁净的海滨沙滩是海鸥馈赠给我们的礼物，为此，我们不仅要心怀感激，而且还要保护他们。

Relish as they might musty bread and mouldy meat, Larie and his mate enjoyed, too, the sport of catching fresh food; and many a clam hunt they had in true gull style. They would fly above the water near the shore, and when they were twenty or thirty feet high, would plunge down head-first. Then they would poke around for a clam, with their heads and necks under water and their wings out and partly unfolded, but not flopping; and a comical sight they were!

尽管拉瑞夫妇很喜欢吃发霉的面包和腐臭的肉，但是，他们也捕捉新鲜食物；一次又一次的捕蛤行动真正体现了海鸥的风范：先在海岸附近的水面上盘旋，飞到二十或三十英尺高时，猛地一头扎进水里，四处戳探，他把脑袋和脖子都伸进水里，而翅膀却露在水面上，半张着，并不拍打，看上去滑稽极了！

After Larie found a clam, he would fly high into the air a hundred feet or so above the rocks, and then, stretching way up with his head, drop the clam from his beak. Easily, with wings fluttering slightly, Larie would follow the clam, floating gracefully, though quickly, down to where it had cracked upon the rocks. The morsel in its broken shell was now ready to eat, for Larie and his mate did not bake their sea-food or make it into chowder. Cold salad flavored with sea-salt was all they needed.

捉到蛤蜊后，拉瑞就会飞到礁石上方一百英尺左右的地方，脖子往上一伸，嘴巴往开一张，把蛤蜊扔下去。然后，轻轻挥动翅膀，拉瑞就能轻而易举地循着蛤蜊落下的足迹，飞到让它粉身碎骨的礁石上，整个过程优雅

After Larie found a clam, he would fly high into the air a hundred feet or so, and then drop it.
捉到蛤蜊后，他就会飞到礁石上方一百英尺左右的地方，把蛤蜊扔下来。

It was not for food alone that Larie and his mate lived that spring.
那年春天,拉瑞和太太活着可不光是为了吃饭。

而又敏捷。碎壳里的美味现在就可以吃了，拉瑞夫妇吃海鲜可不用烤，也不用煮，只要有海盐凉拌过就行。

Exciting as were these hunts with the flocks of screaming gulls, it was not for food alone that Larie and his mate lived that spring. For under the blue of the airy sky there was an ocean, and in that ocean there was an island, and on that island there was a nest, and in that nest there was an egg—the first that the mate of Larie had ever laid. And in that egg was a growing gull, their eldest son—a baby Larie, alone inside his very first world.

和嘎嘎尖叫的海鸥们一起捕捉蛤蜊真是激动人心，然而，那年春天，拉瑞和太太活着可不光是为了吃饭。风轻云淡的碧空下有一片大海，大海里有一处小岛，小岛上有一个鸟巢，鸟巢里有一枚鸟蛋，那是拉瑞太太产的第一枚蛋；蛋里面有一只小海鸥，他就是拉瑞夫妇的长子小拉瑞，他正孤零零地待在他的第一个世界里呢。

3

Peter Piper
矶鹞彼得，小小风笛手

One was named Sandy, because Sandy is a Scotch name and there were blue-bells growing on the rocks; so it seemed right that one of them should have a Scotch name, and what could be better, after all, than Sandy for a sandpiper? One was named Pan, because he piped sweetly among the reeds by the river. One, who came out of his eggshell before his brothers, was named Peter, for his father.

矶鹞的几个孩子中，一个取名沙迪，因为沙迪是苏格兰名字，而周围的岩缝里长满了苏格兰蓝铃花，所以得有一个孩子取个苏格兰名字（注：蓝铃花又称“苏格兰的蓝色铃铛”或“妖精的铃铛”)，再说了，对矶鹞来说，没有比沙迪这个名字更贴切的了(注：sandpiper 是由 sand 和 piper 组成的，sand 意思是沙地)。另一个孩子取名潘，因为他总在河边芦苇林里吹风笛，好听极了。(注：在古希腊神话中，潘是古希腊神话中半人半兽的牧神，也是一位出色的音乐家，能用芦笛吹奏出美妙的曲子，经常吸引山林中的仙女倾听。)比两个弟弟先出壳的老大则随了他爸爸，取名彼得。

But Mother Piper never called her children Sandy and Pan and Peter. She called them all "Pete". She was so used to calling her mate "Pete", that that name was easier than any other for her to say.

不过风笛手妈妈却从未称呼过孩子们“沙迪”、“潘”和“彼得”，她一股脑儿管他们叫“彼迪”。因为叫惯了老公“彼迪”，所以叫孩子们“彼迪”比叫别的什么都顺口。

The three of them played by the river all day long. Each amused himself in his own way and did not bother his brothers, although they did not stray too far apart to talk to one another. This they did by saying,

"Peep", now and then.

三个小家伙整天在河边嬉戏玩耍，各有各的玩法，互不干扰。不过他们不会离得太远，因为三兄弟还要说话呢。他们时不时地会说上一句“彼普”。

About once an hour, and sometimes oftener, Mother Piper came flying over from Faraway Island, crying, "Pete, Pete, Pete," as if she were worried. It is no wonder that she was anxious about Sandy and Peter and Pan, for, to begin with, she had had four fine children, and the very first night they were out of their nest, the darlings, a terrible prowling animal named Tom or Tabby had killed one of her babies.

隔上大约一小时，或者更短时间，风笛手妈妈就会从“远岛”上飞过来，叫着“彼迪，彼迪，彼迪，”一副担心的样子。也难怪她会担心沙迪、潘和彼得，因为，她原本有四个宝宝，都活蹦乱跳的，可是就在他们离巢外出的第一夜，有个宝宝被一个四处觅食的不知是叫雄猫还是虎斑猫的可怕动物给杀死了。

But Peter and Pan and Sandy were too young to

One was named Peter, for his father.
一个孩子随了他爸爸，取名“彼得”。

know much about being afraid. So they played by the river all day long, care–free and happy. Their sweet little voices sounded contented as they said, "Peep," one to another. Their queer little tails looked frisky as they went bob –bob –bob –bing up and down every time they stepped, and sometimes when they didn´t. Their dear little heads went forward and back in a merry sort of jerk. There were so many things to do, and every one of them a pleasure!

可是沙迪、潘和彼得太小，还不知道什么叫害怕。他们整天在河边戏耍，无忧无虑，开心得要命。他们“彼普”、“彼普”地互相招呼，声音细细的、甜甜的，听上去是那样的心满意足。他们每迈一步，奇怪的小尾巴就乒乒乒摇上去，又乒乒乒乒摇下来，有时候又停住不动，看上去活泼极了。可爱的小脑袋一伸一缩，一副兴高采烈的样子。要做的事情太多了，每件都是那么有趣！

Oh! Here was Sandy clambering up the rocky bank, so steep that there was roothold only for the blue –bells, with stems so slender that one name for them is "hair–bell." But Sandy did not fall. He tripped lightly up and about, with sure feet; and where the walking was too hard, he fluttered his wings and flew to an easier place. Once he reached the top of the bank, where the wild roses were blossoming. And wherever he went, and wherever he came, he found good tasty insects to eat; so he had picnic-luncheons all along the way.

哦！瞧！沙迪往石岸上爬呢！太陡了，只有蓝铃花才能扎根落脚；蓝铃花的茎秆非常纤细，所以有人把它们叫作“发丝铃”。可是，沙迪并没有摔下来。他左腾右挪，非常轻巧，脚下一点儿也不犹豫；有些地方太难爬，他就拍着翅膀飞到容易爬的地方。有一次他竟然爬到了盛开着野蔷薇的岸上！无论到何处，也无论来哪里，他都能找到美味的昆虫吃，所以，他总是走一路，野餐一路。

Ho! here was Pan wandering where the river lapped the rocky shore. His long slender legs were just right for wading, and his toes felt comfortable in the cool water. There was a pleasing scent from the sweet gale bushes, which grew almost near enough to the river to go wading, too; and there was a spicy smell when he brushed against the mint, which wore its blossoms in pale purple tufts just above the leaves along the stem. And every now and then, whether he looked at the top of the water or at the rocks on the shore-edge, he found tempting bits of insect game to eat as he waded along.

嗬！瞧，潘正在乱石丛生的岸边溜达呢。河水轻轻地拍着河岸，他那又细又长的双腿正好蹚水；河水凉凉的，脚趾浸在里面舒服极了。美丽的香杨梅丛散发出沁人心脾的芳香，几乎就长在河边，他也可以到那里去蹚水玩。当他从薄荷丛边擦身而过时，一阵辣辣的味道扑鼻而来，原来，薄荷秆上爬满了薄荷叶，上面开满了一簇簇淡紫色的薄荷花。他蹚水的时候，目光会落在水面上，或者岸边的石块上，但无论落在哪里，他都能时不时地找到美味可口的昆虫吃。

Oho! here was Peter on an island as big as an umbrella, with a scooped-out place at one side as deep as the hollow in the palm of a man´s hand. This was shaped exactly right for Peter´s bathtub, and as luck would have it, it was filled to the brim with water. Such a cool splashing—once, twice, thrice, with a long delightful flutter; and then out into the warm sunshine, where the feathers could be puffed out and dried! These were the very first real feathers he had ever had, and he hadn´t had them very long; and my, oh, my! but it was fun running his beak among them, and fixing them all fine, like a grown-up bird. And when he was bathed and dried, there was a snack to eat near by floating toward him on the water.

啊哈！瞧！彼得上“岛”了，那个岛和伞差不多一样大，“岛”的一侧有一块洼地，深浅就像是人的手掌心，形状正好适合给彼得做浴盆，幸好里面还盛满了水。啊，真凉快呀！一下，两下，三下，他兴奋地拍打着翅膀，溅起一身水花。然后，他走出“浴盆”沐浴着温暖的阳光，把身上的羽毛吹开、晒干。这些是他的第一身羽毛，真正的羽毛，刚长出来还没多久呢；哦，天哪！这可真好玩！他像大鸟那样，把嘴巴伸进羽毛里，把它们梳理得漂漂亮亮、干干净净。洗完澡，晾干全身，还有点心可吃，是从附近水面上漂过来的点心。

Oh! Ho! and Oho! It was a day to be gay in, with so many new amusements wherever three brave, fearless little sandpipers might stray.

哦！嗬！啊哈！这日子可真是快活呀！三只勇敢无畏的小矶鹞无论逛到哪里，都有这么多新鲜好玩的事情。

Then came sundown; and in the pleasant twilight Peter and Pan and Sandy somehow found themselves near each other on the bank, still walking forth so brave and bold, and yet each close enough to his brothers to hear a "Peep", were it ever so softly whispered.

太阳下山了。怡人的暮色里，彼得、潘和沙迪不约而同地来到了岸上，他们相依相伴，依然勇往直前，不过，彼此靠得更近了，只要有谁轻轻地低语一声“彼普”，大家都能听得见。

Did it just happen that about that time Mother Piper came flying low over the water from Faraway Island to Nearby Island, calling, "Pete, Pete, Pete," in a different tone, a sort of sundown voice?

怎么这么巧？就在这个时候，风笛手妈妈沿着水面，从“远岛”一路低飞找到了“近岛”，一边飞还一边呼唤：“彼迪，彼迪，彼迪”，语气都变了，是不是一到黄昏，她就是这种语气呢？

Was that the way to speak to three big, most –

grown–up sandpiper sons, who had wandered about so free of will the livelong day?

是不是就该用这种语气呢？三只矶鹞小子都老大不小了，还整天在外面瞎逛游。

Ah, but where were the most–grown–up sons? Gone with the sun at sundown; and, instead, there were three cosy little birds, with their heads still rumpled over with down that was not yet pushed off the ends of their real feathers, and a tassel of down still dangling from the tip of each funny tail.

哎呀！那三个老大不小的小子怎么不见了？太阳落山的时候，和太阳一起没影儿了；现在，瞧，他们又变成了三只居家小宝宝，脑袋乱糟糟的，新羽毛长出来了，可是羽毛根上的绒毛还没褪尽呢；尾巴尖上还耷拉着一绺绒毛穗儿，可真滑稽。

And three dear, sweet, little voices answered, "Peep," every time Mother Piper called, "Pete"; and three little sons tagged obediently after her as she called them from place to place all round and all about Nearby Island, teaching them, perhaps, to make sure there was no Tabby and no Tommy on their camping-ground.

只要风笛手妈妈喊一声“彼迪”，就会听到三声稚嫩的回答：“彼普”，那声音别提多可爱，多甜美了！妈妈喊着他们，从一个地方走到另一个地方，把“近岛”和“近岛”四周走了个遍，每到一处，三个小家伙都顺从地跟在她身后。也许，妈妈是在教他们一定要看看宿营地里有没有“雄猫”和“虎斑猫”吧？

So it was that, after twilight, when darkness was at hand and the curfew sounded for human children to be at home, Peter and Pan and Sandy settled down near each other and near Mother Piper for the night.

果然，黄昏一过，天就黑了，叫人类孩子回家的晚钟也敲响了，彼得、潘和沙迪紧紧地靠在一起，依偎着妈妈准备过夜了。

And where was Peter Piper, who had been abroad the day long, paying little attention to his family? He, too, at nightfall, had come flying low from Faraway Island; and now, with his head tucked behind his wing, was asleep not a rod away from Mother Piper and their three sons.

风笛手爸爸去哪儿了呢?就是那个从早到晚整天都待在外面,很少关心自己家人的爸爸彼得。哦,傍晚时分,他也从“远岛”一路低飞来到了这里,现在,他正把脑袋塞在翅膀后面,在离风笛手妈妈和他们的三个宝贝儿子一箭之地的地方睡觉呢。

Somehow it was very pleasant to know that they were near together through the starlight—the five of them who had wandered forth alone by sunlight.

这一家五口,白天自顾自地到处溜达,夜晚却能聚在一起相依相偎,真是值得我们高兴呀!

But not for long was the snug little Nearby Island to serve for a night camp. Mother Piper had other plans. Like the wise person she was, she let her children find out many things for themselves, though she kept in touch with them from time to time during the day, to satisfy herself that they were safe. And at night she found that they were willing enough to mind what they were told to do, never seeming to bother their heads over the fact that every now and then she led them to a strange camp-ground.

可是，小小的“近岛”虽然温暖舒适，但他们不会在这里长时间夜宿。风笛手妈妈另有打算。她像是有智慧的人类，很多事情都让孩子们自己去领悟，虽然白天她也时不时地和孩子们保持联系，但那也只是为了掌握孩子们的安全状况，好让自己放心。她发现晚上让孩子们做什么，他们都很乐意。虽然每隔一段时间她就带他们换一个陌生的宿营地，但是他们似乎从来都懒得费脑筋去想这件事。

So they did not seem surprised or troubled when, one night soon, Mother Piper, instead of calling them to Nearby Island, as had been her wont, rested patiently in plain sight on a stump near the shore and, with never a word, waited for the sunset hour to reach the time of dusk. Then she flew to the log where Peter Piper had been teetering up and down, and what she said to him I do not know. But a minute later, back she flew, this time rather high overhead, and swooped down toward the little ones with a quick "Pete-weet." After her came Peter Piper flying, also rather high overhead, and swooping down toward his young. Then Mother and Peter Piper went in low, slow flight to Faraway Island.

所以，不久后的一天晚上，风笛手妈妈没有像平常一样喊他们去“近岛”，而是耐心地待在岸边的一根树桩上，一言不发，等待着日落时分薄暮的降临，可他们似乎并不惊讶，也不着急。接着，妈妈飞到正在一根木头上跳上跳下的彼得爸爸跟前，不知对他说了点什么。一分钟之后，她又飞了回来，这次她先飞得高高的，然后短促地叫了一声“彼迪——维迪”，朝下面的小家伙们俯冲下来。她身后跟着彼得爸爸，也是先飞得高高的，然后俯身冲向孩子们。接着妈妈和彼得爸爸一路低飞，慢慢地去“远岛”那边了。

Were they saying good-night to their babies? Were their sons to be left on the bank by themselves, now that they had shaken the last fringe of down from their tails and lost the fluff from their heads? Did they need no older company, now that they looked like grown-up sandpipers except that their vests had no big polka dots splashed over them?

他俩是在向孩子们道晚安吗?是不是因为孩子们已经甩掉了尾巴上那最后一穗绒毛,脑袋上那乱糟糟的绒毛也不见了,就要把他们单独留在岸上呢?是不是因为他们除了背心上还没有洒满大圆点花样以外,看上去已经像大矶鹞了,就不需要大人的陪伴了呢?

Ah, no! At Mother Piper´s "Pete –weet," Peter answered, "Peep," lifted his wings, and flew right past Nearby Island and landed on a rock on Faraway Island. And, "Peep," called Sandy, fluttering after. And, " Peep," said Pan, stopping himself in the midst of his teetering, and flying over Nearby Island on his way to the new camp–ground.

啊,当然不是!听到妈妈“彼迪——维迪”的呼唤,彼得答了一声“彼普”,然后张开翅膀,直接飞过“近岛”,落在了“远岛”的一块岩石上。沙迪也叫了一声“彼普”,拍起翅膀跟了过来。正在玩跷跷板的潘也不玩了,说了一声“彼普”,也越过“近岛”,向新宿营地飞去。

That is how it happened that they had their last luncheon on the shore of Faraway Island before snuggling down to sleep that night.

事情就是这样：那天晚上，他们在“远岛”的岸边加了最后一顿餐，然后就相互依偎着入睡了。

One of the haunts of Peter and Pan and Sandy was Cardinal-Flower Path. This lovely place was along the marshy shore not far from Nearby Island. It was almost white with the fine blooms of water-parsnip, an interesting plant from the top of its blossom head to the lowest of its queer under-water leaves. And here and there, among the lacy white, a stalk of a different sort grew, with red blossoms of a shade so rich that it is called the cardinal flower. Every now and then a ruby-throated hummingbird darted quickly above the water-parsnips straight to the cardinal throat of the other flower, and found refreshment served in frail blossom-ware of the glorious color he loved best of all.

三兄弟经常去一个叫“红花半边莲小径”的地方玩，那里很漂亮，就在“近岛”附近不远处的沼泽沿岸。漂亮的白色水防风花密密麻麻的，整条小路都快变白了。这种植物，从最上头的花尖到最下面浸在水中的奇怪叶子都很有趣。它们像白色的花边，将另一种茎秆植物环绕其间，这种植物的花红艳艳的，所以人们叫它红花半边莲。一只脖颈暗红的蜂鸟时不时地从水防风花上方疾速冲过，直接钻入红花半边莲那红艳艳的花蕊里，那些花蕊就像是一个个花杯，颜色富丽堂皇，是他最喜欢的那种，里面盛着他最喜欢吃的茶点。

And it would be well for all children of men to know that, although three bright active children of sandpipers ran teetering about Cardinal-Flower Path many and many a day, the place was as lovely to look upon at sundown as at sunrise, for not one wonderful spray had been broken from its stem. So it happened, because the children who played there were Sandy and Peter and Pan, that the cardinal flowers lived their life as it was given them by Nature, serving refreshments for hummingbirds through the summer day, and setting seeds according to their kind for other cardinal flowers and other hummingbirds another year.

有一件事情应该让所有人类的孩子都知道，那就是尽管三个活泼可爱的小矶鹞在红花半边莲小径上摇摇晃晃地到处跑了很多很多天，可是，太阳落山的时候，那个地方依然像太阳升起的时候一样美丽怡人，漂亮的花枝都好好地留在茎秆上，三只小矶鹞连一枝都没折。为什么会这样呢？因为在那里玩耍的是矶鹞沙迪、彼得和潘，还因为，红花半边莲的一生都在履行大自然赋予的使命：夏日为鸣蜂端上茶点，并依各自的品种结出各自的种子。这样，来年其他红花半边莲才能发芽开花，其他鸣蜂才有茶点可吃。

But even the charms of Cardinal-Flower Path did not hold Pan and Peter and Sandy many weeks. They seemed to be a sort of gypsy folk, with the love of wandering in their hearts; and it is pleasant to know that, as soon as they were grown enough, there was nothing to prevent their journeying forth with Peter and Mother Piper.

可是“红花半边莲小径”的迷人魅力也无法挽留沙迪、彼得和潘，没过几个星期他们就又要走了。他们就像是吉卜赛人，内心充满了对流浪的渴望；令人高兴的是，他们已经长大，可以和爸爸妈妈一起上路了，没有什么可阻止他们的了。

Of all the strange and wonderful plants and birds and insects they met upon the way I cannot tell you, for, in all my life, I have not traveled so far as these three children went long before they were one year old. They went, in fact, way to the land where the insects live that are so hard and beautiful and gemlike that people sometimes use them for jewels. These are called "Brazilian beetles," and you can tell by that name where the Pipers spent the winter, though it may seem a very far way for a young bird to go, with neither train nor boat to give him a lift.

他们一路上遇到了哪些陌生而奇妙的植物、鸟儿和昆虫呢？这个我可无法告诉你，因为，我一辈子到过的地方还不及这三个小家伙一岁前到过的地方远呢。实际上，他们去了一个遥远的国度，那里的昆虫特别硬，也特别漂亮，就像宝石一样，所以，有时候人们把它们当成珠宝挂在身上，这些昆虫名叫“巴西甲壳虫”，从这个名字你就知道矶鹞一家是在哪里过的冬了。对于一只幼鸟来说，那个地方似乎太远了，因为这一路，既没有火车，也没有轮船可搭。

Not even tired they were, from all accounts, those little feather-folk; and why, indeed, should they be tired? A jaunt from a northern country to Brazil was not too much for a healthy bird, with its sure breath and pure rich blood. There was food enough along the trail—they chose their route wisely enough for that, you may be sure; and they were in no great haste either going or coming.

人人都说，他们甚至都感觉不到累，这些长着羽毛的小家伙！的确，他们为什么要累呢？对一只呼吸正常、热血沸腾的健康鸟儿来说，从北国飞到巴西没有什么大不了的。况且，他们沿途食物供给充足——毫无疑问，旅行路线都是经过了仔细周密地选择的；还有，无论来去，他们都不忙着赶路。

"Coming," did I say? Why, surely! You didn´t think those sandpipers *stayed* in Brazil? What did they care for green gem-like beetles, after all? The only decorations they ever wore were big dark polka dots on their vests. Perhaps they were all pleased with them, when their old travel-worn feathers dropped out and new ones came in. Who can tell? They had a way of running their bills through their plumage after a bath, as if they liked to comb their pretty feathers.

"来"？我刚才说"来"了吗？当然啦！你总不会以为矶鹞们就一直"留守"巴西不回来了吧？毕竟，那些像绿宝石一样的甲壳虫关他们什么事呢？他们佩戴过的唯一装饰品就是背心上圆圆的大黑点。一路风餐露宿、饱经风吹雨打，身上的羽毛变得破旧不堪，纷纷脱落，取而代之的是新羽毛，上面还点缀着这些圆圆的大黑点，你说他们能不喜欢吗？谁知道呢？每次洗完澡，他们都自有办法把嘴巴伸到羽衣里，仔细梳理那些漂亮的羽毛，一副乐此不疲的样子。

Be that as it may, there was something beneath their feathers that quickened like the heart of a journeying gypsy when, with nodding heads and teetering tails, they started again for the north.

别看他们那么悠然自得，可是，当他们点着头、摇动尾巴北归的时候，那身羽毛下面的某个东西也像旅途中吉卜赛人的心脏那样加速跳动起来。

Did they dream of a bank where the blue-bells grew, and a shore spiced with the fragrance of wild mint?

他们是不是梦到了开满蓝铃花的河堤？是不是梦到了到处飘散着野薄荷香的湖畔？

No one will ever know just how Nature whispers to the bird, "Northward ho!" But we know they come in the springtime, and right glad are we to hear their voices.

没人知道大自然是怎样悄声低语地叫鸟儿们“向北走”的。可是，我们知道，一到春天他们就回来，我们知道，听到他们的鸣叫我们有多高兴！

So Peter Piper, Junior, came back again to the shore of Nearby Island. And do you think Sandy and Pan walked behind him for company, calling, "Peep," one to another? And do you think Mother Piper and Father Peter showed him the way to Faraway Island at sun-down, and guarded him o´ nights? Not they! They were busy, every one, with their own affairs, and Peter would just have to get along without them.

小彼得又回到了“近岛”，来到了岸边。你以为沙迪和潘还会跟在他后面陪着他，一边走还一边彼此喊着“彼普”？你以为太阳下山的时候，风笛手妈妈和彼得爸爸为他指了去“远岛”的路，晚上还守护在他身边？他们才不呢！大家都各自忙各自的事情呢，彼得只能单独行动了。

Well, Peter could—Peter and Dot. For of course he was a grown-up sandpiper now, with a mate of his own, nodding her wise little head the livelong day, and teetering for joy all over the rocks where the red columbine grew.

不过,彼得能够自己行动了——他有多蒂呢。当然了，彼得现在已经是一只大矶鹞了，有自己的伴侣了。多蒂整天都点着聪明的小脑袋,在长满红色耧斗草的岩石堆上摇摇晃晃地走来走去,过得很开心。

The spot she teetered to most of all was a little cup-shaped hollow high up on the border of the ledge, where the sumachs were big as small trees and where the sweet fern scented the air. The hollow was lined tidily and softly with dried grass, and made a comfortable place to sit, no doubt. At least, Dot liked it; and Peter must have had some fondness for it, too, for he slipped on when Dot was not there herself. It just fitted their little bodies, and there were four eggs in it of which any sandpiper might well have been proud; for they were much, much bigger than most birds the size of Dot could ever lay. In fact, her little body could hardly have covered them snugly enough to keep them

The spot she teetered to most of all.
她摇摇晃晃经常光顾的地方。

warm if they had not been packed just so, with the pointed ends pushed down into the middle of the rather deep nest.

她最常去的地方是一块形状像茶杯的小洼地,在凸岩边上,那里漆树长得像小树一样大,空气中弥漫着香蕨木的芬芳。小洼地里面铺着整整齐齐、软软乎乎的干草,坐进去肯定很舒服,反正多蒂喜欢这里;彼得肯定也挺喜欢的,因为多蒂不在里面时,他就会溜进来坐上去。小洼地刚好能容下他们的小身子,里面还有四枚蛋呢,四枚让任何矶鹞都为之自豪的蛋;这几个蛋很大,比大多数和多蒂个头一样大的鸟产下的蛋大多了。其实,如果不是他们尖头朝下挤进深深的巢心的话,就凭多蒂那小身板,根本就不可能把他们舒舒服服地罩起来,为他们保温。

The eggs were creamy white, with brown spots splashed over them—the proper sort of eggs (if only they had been smaller) to tuck beneath a warm breast decorated with pretty polka dots. But still, they must have been her very own, or Dot could not have taken such good care of them.

这几枚蛋白得像奶油,上面洒着褐色的斑点——裹在多蒂那长满漂亮圆点的前胸下，又温暖又般配（如果蛋再小一点儿的话就更相配了）。不过，无论如何，这些蛋肯定得是她自己的，要不多蒂才不会这么精心地照顾他们呢。

Because of this care, day by day the little body inside each shell grew from the wonderful single cell it started life with, to a many-celled creature, all fitted out with lungs and a heart and rich warm blood, and very slender legs, and very dear heads with very bright eyes, and all the other parts it takes to make a bird. When the birds were all made, they broke the shells and pushed aside the pieces. And four more capable little rascals never were hatched.

在她的精心呵护下，蛋壳里的小生命一天天长大了：最初只是一个神奇的单细胞，后来变成了一只完完整整的多细胞鸟儿，不仅有肺、有心、有涌动的热血，还有细细的腿和可爱的脑袋，脑袋上还有分外明亮的眼睛。等大家都变成了鸟儿，他们就开始敲击蛋壳，把敲下来的碎片拨拉到一边，于是，四只能耐无比的小淘气鬼就孵出来了。

Why, almost before one would think they had had time to dry their down and stretch their legs and get used to being outside of shells instead of inside, those little babies walked way to the edge of the river, and from that time forth never needed their nest.

咦，这四个小宝宝怎么这么快就弄干了身上的绒毛？好像连腿都没伸一下就适应了蛋壳外面的世界？只见他们一路远行来到了河边，从此，就再也没有回过他们的小巢。

And look! The fluffy, cunning little dears are nodding their heads and teetering their tails! Yes, that proves that they must be sandpipers, even if we did have doubts of those eggs. Ah! Dot knew what she was about all along. The size of her eggs might fool a person, but she had not worried. Why, indeed, should she be troubled? Those big shells had held food-material enough, so that her young, when hatched, were so strong and well-developed that they could go wandering forth at once. They did not lie huddled in their nest, helplessly begging Peter Piper and Mother Dot to bring them food. Not they! Out they toddled, teetering along the shore, having picnics from the first—the little gypsy babies!

瞧！这些毛茸茸的小家伙们正在点头、摇尾巴呢，好小巧、好可爱呀！是呀，这足以证明他们就是矶鹞，就算我们对那四枚蛋有疑问也没关系了。哈！多蒂一直都很清楚她在忙什么。她的蛋很大，也许会让人类迷惑，但是，她可从来不会担心弄错。

嗨，的确，她有什么好担心的呢？那些大大的蛋壳里面装着足够可吃的东西，孩子们孵出来的时候，身体结实，发育良好，马上就能出去溜达。他们才不会蜷缩在窝里，可怜兮兮地向爸爸小彼得和妈妈多蒂要吃的呢。这可不是他们的风格！从一开始，他们就迈着蹒跚的脚步离开蛋壳，摇摇晃晃地来到岸边，四处野餐——简直就是一群吉卜赛小宝宝！

Tabby did not catch any of them, though one night she tried, and gave Dot an awful scare. It was while they were still tiny enough to be tucked under their mother's feathers after sundown, and before they could manage to get, stone by stone, to Nearby Island. So they were camped on the shore, and the prowling cat came very near. So near, in fact, that Mother Dot fluttered away from her young, calling back to them, in a language they understood, to scatter a bit, and then lie so still that not even the green eyes of the cat could see a motion. The four little Pipers obeyed. Not one of them questioned, "Why, Mother?" or whined, "I don´t want to," or whimpered, "I´m frightened," or boasted, "Pooh, there´s nothing here."

虎斑猫一直想抓他们，可是从未得逞，尽管一天晚上她差点得手，着实把多蒂妈妈吓坏了。那时他们还很小，太阳下山之后全都能挤在妈妈身下藏起来；那时，他们还没有越过一块块岩石到达“近岛”呢。当时他们正在河边露营，那只大猫悄悄地过来潜伏在附近，实际上，她离得太近了，结果，多蒂妈妈一下子扑棱起翅膀从孩子们身边飞开，一边飞一边回头用孩子们能听懂的语言叫他们散开、趴下别动。他们真的就一动不动地趴在那里，猫的眼睛再贪婪也看不到一丁点儿动静。风笛手家的四个小家伙都很听话，谁也没问“为什么呀，妈妈”，或者小声嘀咕“我不”，谁也没哭哭啼啼地哼哼“我怕”，或者不以为然地犟嘴“切！根本没什么嘛”。

Dot led the crouching enemy away by fluttering as if she had a broken wing, and she called for help with all the agony of her mother–love. "Pete," she cried, "Pete," and "Pete, Pete, Pete! "

多蒂装作翅膀受伤的样子拍打着翅膀，引开了潜伏的敌人，与此同时，母爱的力量让她拼尽全力大声呼救："彼迪！""彼迪，彼迪！""彼迪，彼迪，彼迪！"

No one who hears the wail of a frightened sandpiper begging protection for her young can sit unmoved.

无论是谁，只要听到矶鹞惊恐万状的求救声，都不会坐视不管的。

Someone at the Ledge House heard Dot, and gave a low whistle and a quick command. Then there was a dashing rush through the bushes, that sounded as if a dog were chasing a cat. A few minutes later Dot´s voice again called in the dark—this time, not in anguish of heart, but very cosily and gently. "Pete–weet?" she whispered; and four precious little babies murmured, "Peep," as they snuggled close to the spotted breast of their mother.

凸岩屋那边有人听到了多蒂的求救声，他一声低鸣，冲上半空俯瞰片刻，然后，就以迅雷不及掩耳之势嗖嗖地冲过灌木丛，那声音就好像是一只狗在猛追一只猫。几分钟后，夜幕中又传来多蒂的声音，不过这次不是惊恐的求救声，而是温柔亲切的低声问候"彼迪——维迪？"四个小宝贝也一边往妈妈那长满斑点的胸脯下面挤，一边喃喃地回答道："彼普。"

So it happened that two sons and two daughters of Peter Piper, Junior, played and picnicked and bathed by the river. The one who had first pipped his eggshell was named Peter the Third, for his father and his grandfather, and a finer young sandpiper never shook the fluff of down from his head or the fringe from his tail, when his real feathers pushed into their places.

就这样，风笛手小彼得的两个儿子两个女儿在河边嬉戏玩耍，一会儿野餐，一会儿又洗澡。头一个啄破蛋壳的那个孩子名字随他的爷爷和爸爸，叫彼得三世，他比爷爷和爸爸还漂亮，真正的羽毛刚刚长出来的时候，脑袋上的绒毛依然毛茸茸的，尾巴上的绒毛穗也仍旧耷拉着。

What his brother and sisters were named, I never knew; and it didn´t matter much, for their mother called them all "Pete".

我一直不知道他的弟弟和两个妹妹叫什么;不过这没多大关系,因为他们的妈妈管他们都叫“彼迪”。

Peter the Third and the others grew up as Pan and Peter and Sandy had grown, dallying happily along the river–edge, and as happily accepting the guidance of their mother, who made her slow flight from Faraway Island every now and then, usually so low that her spotted breast was reflected in the clear water as she came, the white markings in her wings showing above and below.

当年潘、彼得和沙迪是怎么长大的,彼得三世兄妹就是怎么长大的,他们快快乐乐地在河边嬉戏玩耍,高高兴兴地听妈妈的教导。妈妈每隔一段时间就从“远岛”慢慢飞过来,通常,她都飞得很低,所以,她每次来的时候,清澈的河水里都会倒映着她那洒满斑点的胸脯,翅膀上面和翅膀下面的白色标记一览无余。

Dallying happily along the river–edge.
在河边快乐地嬉戏玩耍。

Of course, as soon as the season came for their migration journey, the four of them started cheerfully off with Peter and Dot, for a leisurely little flight to Brazil and back—to fill the days, as it were, with pleasant wanderings, from the time the hummingbird fed at the feast of the cardinal flower in late summer, until he should be hovering over the columbine in the spring.

当然,一旦迁徙季节到来,四个小家伙就会兴高采烈地跟着彼得和多蒂开始短途旅行,他们不慌不忙地飞到巴西,然后又不慌不忙地飞回来。从夏末鸣蜂在红花半边莲上美餐那天开始,一直到春天他在耧斗菜上盘旋期间,每天他们都四处溜达,别提多快活了!

4

Gavia of Immer Lake
爱默湖上的潜鸟加菲娅

Once upon a time, it was four millions of years ago. There were no people then all the way from Florida to Alaska. There was, indeed, in all this distance, no land to walk upon, except islands in the west where the Rocky Mountains are now. That is the only place where the country that is now the United States of America stuck up out of the water. Everywhere else were the waves of the sea. There were no people, even on the Rocky Mountain Islands. None at all.

很久以前，也就是四百万年前，佛罗里达到阿拉斯加沿途荒无人烟，除了西边有一片岛屿之外，根本就没有陆地可走，那片岛屿所在的地方就是现在的落基山脉。今天美国所处的这片陆地，当时只有那一块地方是露出水面的，其余地方都是一片汪洋大海。即使是落基山岛上也渺无人烟，一点儿都没有！

No, the creatures that visited those island shores in those old days were not people, but birds. Nearly as large as men they were, and they had teeth on their long slender jaws, and they had no wings. They came to the islands, perhaps, only at nesting-time; for their legs and feet were fitted for swimming and not walking, and they lived upon fish in the sea. So they dwelt, with no man to see them, on the water that stretched from sea to sea; and what their voices were like, no man knows.

是的，很久、很久以前，光顾那些岛屿的不是人类，而是鸟类。它们的体型和人类一样大，又长又细的嘴巴里长着牙齿，它们没有翅膀。或许它们只在筑巢期间才会来到岛上，因为它们的腿和脚只适合游泳，不适合走路，它们以海里的鱼类为食。它们居住的水域，延伸出去就是一片又一片的大海，没有人类看到它们；至于它们的声音是什么样的，也没人知道。

A million years, perhaps, passed by, and then another million, and maybe another million still; and the birds without wings and with teeth were no more. In their places were other birds, much smaller—birds with wings and no teeth; but something like them, for all that: for their feet also were fitted for swimming and not walking, and they, too, visited the shore little, if at all, except at nesting-time, and they lived upon fish in the water.

也许是过了一百万年，接着又过了一百万年，也许又过了第三个一百万年，那些没有翅膀却有牙齿的鸟消失了。它们生活过的地方生活着另一种鸟类，体型要小得多，有翅膀，却没有牙齿；不过虽然如此，还是有很多相似的地方：它们的脚也只适合游泳，不适合走路；如果不是筑巢季节，它们极少光顾那些岛屿；另外，它们也以水里的鱼类为食。

And what their voices were like, all men may know who will go to the wilderness lakes and listen; for, wonderful as it may seem, these second birds have come down to us through perhaps a million years, and live to-day, giving a strange clear cry before a storm, and at other times calling weirdly in lone places, so that men who are within hearing always say, "The loons are laughing".

那么，它们的叫声是怎样的呢？只要到那没有人烟的湖泊去听听，就知道了，因为尽管这事有点神奇，但是，这第二类鸟的确是经过了大约一百万年才来到我们身边，一直生活到如今的。暴风雨来临时，它们会发出奇怪、清脆的尖叫；平时，在形单影只的时候，它们的叫声也非常古怪，听到的人总是说："听，潜鸟笑了。"

Gavia was a loon who had spent the winter of 1919–1920 on the Atlantic Ocean. There had hardly been, perhaps, in a million years a handsomer loon afloat on any sea. Even in her winter coat she was beautiful; and when she put on her spring suit, she was lovelier still.

加菲娅是一只潜鸟，1919 到 1920 年的那个冬天，她是在大西洋上度过的。或许，她是一百万年以来，在所有海域上漂浮过的最漂亮的潜鸟了。即使是穿着冬衣，她也很漂亮；等换上春装，那就更加美丽动人了。

She and her mate had enjoyed the sea–fishing and

had joined a company of forty for swimming parties and other loon festivities; for life on the ocean waves has many interests, and there is never a lack of entertainment. The salt-water bathing, diving, and such other activities as the sea affords, were pleasant for them all. Then, too, the winter months made a chance for rest, a change from home-duties, and a freedom from looking out for the children, that gave the loons a care-free manner as they rode the waves far out at sea.

她和她的伴侣喜欢在海里捕鱼，他们还加入了一个有四十个同伴的团队，经常出席游泳晚会和其他欢庆活动。茫茫大海，波涛汹涌，生活的乐趣很多，从来不乏娱乐活动。大伙儿最喜欢咸水浴和潜水，还有其他这样的大海为他们举办的活动。然后，到了冬天，也很惬意。他们刚好能利用那几个月好好休息一下，不用再做家务，也不用再照看孩子，所以，当潜鸟们乘着海浪漂到大海深处去的时候，那神情别提多悠闲，多自在了！

Considering all this, it seems strange, does it not, that when the spring of 1920 had gone no further than to melt the ice in the northern lakes, Gavia and her mate left the sea and took strong flight inland.

Immer Lake
爱默湖

既然如此，事情是不是就有点奇怪了？1920 年的春天才刚刚开始，北方湖里的冰还没解冻呢，加菲娅和她的伴侣就离开了大海，义无反顾地往内陆飞去了。

What made them go, I cannot explain. I do not understand it well enough. I do not really know what urges the salmon to leave the Atlantic Ocean in the spring and travel up the Penobscot or the St. John River. I never felt quite sure why Peter Piper left Brazil for the shore where the blue-bells nod. All I can tell you about it is that a feeling came over the loons that is called a migration instinct; and, almost before Gavia and her mate knew what was happening to them, they had flown far and far from the Ocean, and were laughing weirdly over the cold waters of Immer Lake.

他们为什么要离开大海而飞往内陆？这个我无法解释。我也不是很明白。我真的不是很清楚为什么大马哈鱼一到春季就离开大西洋，游往佩诺布斯科特河或圣约翰河。我从来都不十分肯定，为什么矶鹞彼得要离开巴西，飞往那个开满蓝铃花的海滨。我只知道，潜鸟突然就有了一种感觉，这种感觉叫作“迁徙直觉”。所以，加菲娅和她的伴侣几乎还没明白是怎么回事的时候，就已经远远地离开了海洋，来到了爱默湖，冲着冰冷的湖水奇怪地大笑起来了。

The shore was dark with the deep green of fir trees, whose straight trunks had blisters on them where drops of fragrant balsam lay hidden in the bark. And here and there trees with white slender trunks leaned out over the water, and the bark on these peeled up like pieces of thin and pretty paper. Three wonderful vines trailed through the woodland, and each in its season blossomed into pink and fragrant bells. But what these were, and how they looked, is not a part of this story, for Gavia never wandered among them. Her summer paths lay upon and under the water of the lake, as her winter trails had been upon and under the water of the sea.

湖滨的杉树郁郁葱葱，遮天蔽日。笔直的树干上有很多水泡，水泡那里的树皮里面藏着一滴滴的香脂。有一种树，树干白白的、细细的，探出身子映在湖面上，上面皲裂的树皮，就像是一块块漂亮的薄纸片。三株藤萝在林地里蔓延生长，季节一到，每一株都挂满香气四溢的粉红色小铃铛。不过，这是什么花，长得什么样和本故事无关，因为加菲娅从未拜访过它们。夏天，她只在湖水的水面和水下活动，就好像冬天她只在海水的水面和水下活动一样。

Ah, if she loved the water so, why did she suddenly begin to stay out of it? If she delighted so in swimming and diving and chasing wild wing-races over the surface, why did she spend the day quietly in one place?

啊，既然她那么喜欢水，为什么却突然要离开呢？既然她那么喜欢游泳和潜水，那么喜欢在水面上撒欢、进行飞行比赛，为什么却整天静静地待在一个地方呢？

Of course you have guessed it! Gavia was on her nest. She had hidden her two babies among the bulrushes for safety, and must stay there herself to keep them warm. They were not yet out of their eggshells, so the only care they needed for many a long day and night was constant warmth enough for growth. They lay near each other, the two big eggs, of a color that some might call brown and some might call green, with dark-brown spots splashed over them.

哈，你猜对了！加菲娅正在巢里孵蛋呢。为了安全，

Two babies, not yet out of their eggshells, hidden among the rushes.
两个还未出壳的宝宝藏在灯芯草里。

她把两个小宝宝藏在了灯芯草里，为了给他们保暖，她留了下来。宝宝还没出壳，为了发育成长，在接下来的很多个日日夜夜里，他们需要的唯一照顾就是保持一定的温度。这两枚蛋大大的，紧紧地挨在一起，颜色有人说是棕的，也有人说是绿的，上面洒满了深褐色的斑点。

The nest Gavia and her mate had prepared for them was a heap of old wet reeds and other dead water–plants, which they had piled up among the stems of the rushes until it reached six inches or more out of the water. They were really in the centre of a nest island, with water all about them. So, you see, Gavia was within splashing distance of her fishing–pool after all.

加菲娅夫妇为两个宝宝筑的巢是用湿湿的枯芦苇和别的一些枯水草堆成的，架在灯芯草的茎秆之间，高出水面六英寸多（大约 15 厘米）。所以，这个巢就像是一个四面环水的小岛，两个小宝宝就躺在小岛的中心。所以你瞧，加菲娅离她的鱼池也只不过是一步之遥嘛。

She and her mate, indeed, were in the habit of

making their nests here in the cove; though the two pairs of Neighbor Loons, who built year after year farther up the lake, chose places on the island near the water –line in the spring; and when the water sank lower later on, they were left high and dry where they had to flounder back and forth to and from the nest, as awkward on land as they were graceful in the water.

其实，他们夫妇就习惯在这样的水湾处筑巢；而他们的邻居，常年住在湖上游那边的那两对潜鸟夫妇，春天的时候，他们总是把巢筑在岛上离水面很近的地方，后来水面下降，他们就陷入了困境，为了觅食，不得不在他们的巢和湖水之间一趟一趟地辗转往返，他们在水面上有多优雅，在陆地上就有多不自在。

Faithful to her unhatched young as Gavia was, it is not likely that she alone kept them warm for nearly thirty days and nights; for Father Loon remained close at hand, and would he not help her with this task?

尽管加菲娅对她那两个还未孵化的宝宝尽职尽责，但要在近三十个日日夜夜的时间里仅靠她自己为他们保温是不可能的；潜鸟爸爸就在跟前，难道他不帮忙吗？

Gavia, sitting on her nest, did not look like herself of the early winter months when she had played among the ocean waves. For her head and neck were now a beautiful green, and she wore two white striped collars, while the back of her feather coat was neatly checked off with little white squarish spots. Father Loon wore the same style that she did. Summer and winter, they dressed alike.

正在巢里孵小宝宝的加菲娅,看起来一点儿都不像初冬时节乘着海浪踏着波涛嬉戏玩耍时的加菲娅。现在,她的头和脖颈都变成了漂亮的绿色,还戴着两个白条纹项圈,羽衣的后背上是整整齐齐的方形小斑点。潜鸟爸爸的穿戴和她一样。无论夏天还是冬天,他们的穿戴都一样。

Yes, a handsome couple, indeed, waited that long month for the birth of their twins, growing all this time inside those two strong eggshells. At last, however, the nest held the two babies, all feathered with down from the very first, black on their backs and gray shading into white beneath.

是的,的确是一对漂亮的小夫妻,在那一个月漫长的等待中,他们期待着双胞胎宝宝的诞生;与此同时,两个小宝宝也正在硬硬的蛋壳里面不停地长呢。不管怎样，最终鸟巢盛上了两个小宝宝，他们一出壳浑身就长满了小羽绒，后背黑黑的，往下慢慢变浅，最后到肚子底下就成白的了。

Did I say the nest held them? Well, so it did for a few hours. After that, they swam the waters of Immer Lake, and their nest was home no longer. Peter Piper´s children themselves were not quicker to run than Gavia´s twins were to swim and dive.

我说的是不是鸟巢"盛"上了两个小宝宝？嗨，只盛了几个小时而已。这几个小时一过，他们就已经在拨弄爱默湖的湖水了，再也不把鸟巢当成自己的家了。矶鹞彼得的孩子们出生后很快就能出去跑了，可是，还是比不上潜鸟加菲娅的双胞胎学游泳和学潜水的速度快。

I think, perhaps, they were named Olair; for Gavia often spoke in a very soft mellow tone, saying, "Olair"; and her voice, though a bit sad, had a pleasing sound. So we will call them the two Olairs.

我觉得，也许两个宝宝的名字叫奥莱尔，因为加菲娅常常说“奥莱尔”，语调又温柔，又圆润，尽管略带忧伤，不过很好听。所以我们就把两个宝宝叫“双胞胎奥莱尔”吧。

They were darlings, those baby loons, swimming about (though not very fast at first), and diving out of sight in the water every now and then (but not staying under very long at the beginning). Then, when they were tired or in a hurry, they would ride on the backs of Gavia and Father Loon: and they liked it fine, sailing over the water with no trouble at all, just as if they were in a boat, with someone else to do the rowing.

这些潜鸟宝宝好可爱呀！他们在水中游来游去（尽管刚开始不是很快），还时不时地潜入水中藏起来（可是，刚开始不会在水下待很久）。然后，累了或急了，就趴在加菲娅妈妈和潜鸟爸爸的背上让他们背着走。他们俩很喜欢这样无忧无虑地在水面上航行，就好像是坐在一个小船里，自有别人替他们划船。

Oh, yes, they were darlings! Had you seen one of them, you could hardly have helped wanting to cuddle him. But do you think you could catch one, even the youngest? Not a bit of it. If you had given chase in a boat, the wee-est loon would have sailed off faster yet on the back of his father; and when you grew tired and stopped, you would have heard, as if mocking you, the old bird give, in a laughing voice, the *Tremble Song*:

"O, ha-ha-ha, ho! —O, ha-ha-ha, ho! —O, ha-ha-ha, ho! —O, ha-ha-ha, ho! —"

哦，是的，他们太可爱了！他俩无论是谁，如果你看见，都会忍不住地想去抱他。可是，你以为你能逮住小潜鸟吗？哪怕是最小的那只？你想都别想了。假如你乘船去追，最小的那只会坐在爸爸的背上比你跑得还快；当你累得划不动时，就会听见大潜鸟唱的《颤歌》，声音笑嘻嘻的，就像是在嘲笑你：

“哦，哈——哈哈，嗬！——哦，哈——哈哈，嗬！——哦，哈——哈哈，嗬！——哦，哈——哈哈，嗬！——”

If you had tried again a few days later, the young loon would have been able to dive and swim by himself out of sight under water, the old ones giving him warning of danger and telling him what to do.

要是你过几天再试，那时候小潜鸟就已经能自己潜入水下，游到你看不见的地方去了。大潜鸟只管向他发送危险信号，告诉他该怎么做。

But no child chased the two Olairs and no law-breaker fired a shot at Gavia or Father Loon. They had frights and narrow escapes in plenty without that; but those were of the sorts that loons get used to century after century, and not modern disasters, like guns, that people have recently brought into wild places. For the only man who dwelt on the shore of Immer Lake was a minister.

不过没有孩子追逐双胞胎奥莱尔，也没有罪犯向加菲娅或者潜鸟爸爸开枪。即便如此，他们也常常受到惊吓，多少次都是死里逃生。可那都是多少世纪以来潜鸟们已经习以为常的遭遇，不像枪支那样，是人类新近带给野生世界的现代灾难。这里的平静都是因为爱默湖畔的唯一居民是个牧师。

Because he loved his fellow men, this minister of Immer Lake spent part of his days among them, doing such service to the weak of spirit as only a minister can do, who has faith that there is some good in every person. At such times he was a sort of servant to all who needed him.

因为深爱着他的人类同胞，爱默湖畔的这个牧师有一部分时间和人们在一起，他像牧师那样，安慰虚弱的灵魂，因为只有牧师相信每个人都有善的一面。每当此时，他就像是一个神的使者，为所有需要他的人提供帮助。

Because he loved, also, his fellow creatures who had lived in the beautiful wild places of this land much longer than any man whatsoever, he spent part of his days among them. At such times he was a sort of hermit.

因为他也深爱着他的鸟类同胞，深爱着生活在这片美丽野生世界里最早的居民，所以，他有一部分时间和鸟儿在一起。每当此时，他就好像是个隐士。

Then no handy trolley rumbled by to take him on his near way. No train shrieked its departure to distant places where he might go. There was no interesting roar of mill or factory making things to use. There was no sociable tread of feet upon the pavement, to give him a feeling of human companionship.

可是,没有隆隆的电车能让他方便搭乘;没有鸣着汽笛出发的火车载他奔向远方;没有忙着制造用品的工厂那热闹的轰鸣;人行道上也没有熟悉亲切的脚步声能让他感受到人类的温暖。

But, for all that, it was not a silent world the minister found at Immer Lake. On sunny days the waves, touching the rocks on the shore, sang gently, "Bippo–bappo, bippo–bappo." The trees clapped their leaves together as the breezes bade them. The woodpeckers tapped tunes to each other on their hollow wooden drums. The squirrels chattered among the branches. At dawn and at dusk the thrushes made melodies everywhere about.

但是,即便如此,牧师发现爱默湖并不是一个寂静无声的世界。阳光明媚的日子里,波涛"噼啪——噼啪,噼啪——噼啪"地拍打着湖边的岩石,就像是在轻轻地歌唱;微风吹过树林,树叶哗哗地拍起了巴掌;啄木鸟你应我和地敲着他们的空心木鼓;树枝上,松鼠吱吱吱地说个没完没了；黄昏和清晨，到处都是画眉鸟那美妙的歌声。

On stormy nights the waves slapped loudly upon the rocks. The branches whacked against one another at the mighty will of the wind. The thunder roared applause at the fireworks the lightning made. And best of all, like the very spirit of the wild event, there rang the strange, sweet moaning *Storm Song of the Loon*:—

"A–a–ah l–u–u–u–u–u–u´ la. A–a–ah l–u–u–u–u–u–u´ la. A–a–ah l–u–u–u–u–u–u´ la. A–a–ah l–u–u–u–u–u–u´ la."

暴风雨之夜，波涛汹涌，拍打着岩石，发出巨响；怒吼的狂风中，树枝摇曳，相互撞击；隆隆的雷鸣仿佛是在为耀眼的闪电欢呼叫好；最精彩的莫过于响彻夜空的《潜鸟之暴风雨之歌》了，歌声奇特，既悦耳又悲凉，那才是真正的荒野之歌呀！

"啊——啊——哈，噜——呜——呜——呜——呜——呜——呜，啦。啊——啊——哈，噜——呜——呜——呜——呜——呜——呜，啦。啊——啊——哈，噜——呜——呜——呜——呜——呜——呜，啦。"

The minister of Immer Lake liked that song, and

he liked the other music that they made. So it was that he sat before his door through many a summer twilight, and played on his violin until the loons answered with the *Tremble Song*:—

"O, ha–ha–ha, ho! O, ha–ha–ha, ho! O, ha–ha–ha, ho! O, ha–ha–ha, ho! "

爱默湖畔的牧师喜欢这支歌，他也喜欢潜鸟唱的其他歌。所以，夏日里，许许多多个黎明薄暮中，他就坐在门前，拉着小提琴，等着潜鸟唱起《颤歌》来应和他的琴声：

"哦，哈——哈哈，嗬！哦，哈——哈哈，嗬！——哦，哈——哈哈，嗬！——哦，哈——哈哈，嗬！"

Then they would swim up and up, until they floated close to his cottage, feeding unafraid near by, while he played softly.

接着，潜鸟们就会一直往上游呀游，一直游到牧师的小木屋附近，一边听牧师轻柔地拉琴，一边在那里觅食，一点儿都不害怕。

Often, when Gavia and her mate were resting

there or farther up the lake, some other loon would fly over; and then Father Loon would throw his head way forward and give another sort of song. "Oh–a–lee′! " he would begin, with his bill wide open; and then, nearly closing his mouth, he would sing, "Cleo′–pe′′–a–rit′." The "Oh" starts low and then rises in a long, drawn way. Perhaps in all the music of Immer Lake there is nothing queerer than the *Silly Song of Father Loon*:—

"Oh–a–lee′! Cleo′–p′′–a–rit′, cleo′–pe′′–a–rit′, cleo′–per′′–wer–wer! Oh–a–lee′! Cleo′–p′′–a–rit′, cleo′–pe′′–a–rit′, cleo′–pe′′–wer–wer! "

加菲娅夫妇在湖畔或湖水深处休息时，常常会有一些潜鸟从头顶飞过，于是，潜鸟爸爸就会把脑袋伸得长长的，唱起另一种歌。刚开始他会张大嘴巴，唱声"啊噢——啊——哩'！"然后，合拢嘴巴，几乎是闭着嘴唱："克莱欧——呸——啊——瑞特。"唱"啊噢"的时候，开始是低音，然后，就拖得长长的升成了高音。或许爱默湖的所有歌声中，再也找不出比《潜鸟爸爸的笨歌》更奇特的歌了：

"啊噢——啊——哩！ 克莱欧——呸——啊——瑞特，克莱欧——呸——啊——瑞特，克莱欧——呸——喔尔——喔尔！ 克莱欧——呸——啊——瑞特，克莱欧——呸——喔尔——喔尔！"

Such were the songs the two Olairs heard often and again, while they were growing up; and they must have added much to the interest of their first summer.

双胞胎奥莱尔小时候经常听、反复听的就是这样的歌，这些歌一定给他们那兴致盎然的第一个夏季增添了许多乐趣。

Altogether they had endless pleasures, and were as much at ease in the water as if there were no more land near them than there had been near those other young birds that had teeth and no wings, four million years or so ago. Their own wings were still small and flipper–like when, about the first of August, they were spending the day, as they often did, in a small cove. They were now about two –thirds grown, and their feathers were white beneath and soft bright brown above, with bars of white spots at their shoulders. They had funny stiff little tails, which they stuck up out of the water or poked out of sight, as they wished. They swam about in circles, and preened their feathers with their bills, which were still small and gray, and not black like those of the old birds.

总之，他们其乐无穷。和大约四百万年前那些有牙没翅的小鸟一样，他们在水里悠然自得、轻松自在，就好像他们身旁也没有什么陆地似的。八月一日前后，他们像往常一样在一个小水湾里打发日子，他们的翅膀依然很小，像鱼鳍一样。现在他俩已经发育了三分之二，腹部的羽毛白白的，背部是柔和、光亮的褐色，肩部是一条条的白点。他们的小尾巴硬硬的，可好玩了，可以随心所欲地戳出水面，或者插入水下藏起来。他们转着圈游来游去，还用嘴巴梳理身上的羽毛，他们的嘴巴还小，是灰色的，不像那些大潜鸟的嘴巴是黑色的。

After a time Gavia came swimming toward them, all under water except her head. Suddenly Father Loon joined her, and they both began diving and catching little fishes for the two Olairs. For the vegetable part of their dinner they had shreds of some water plant, which Gavia brought them, dangling from her bill. Surely never a fresher meal was served than fish just caught and greens just pulled! No wonder it was that the young loons grew fast, and were well and strong. After the twins were fed, Gavia and Father Loon sank from sight under the water, heads and all, and the O-lairs saw no more of them for two hours or so, though they heard them now and then singing, sometimes the

Tremble Song and sometimes the *Silly Song*.

过了一会儿，加菲娅往他们这边游来了，她全身都没在水下，只有脑袋露在外面。突然潜鸟爸爸也赶来了，于是，他们双双开始潜水，为两个奥莱尔捉小鱼吃，因为晚餐素菜他们只吃了一些妈妈用嘴巴衔来的碎条水草。当然了，再也没有比刚捉的鱼、刚拽的草更鲜美的饭菜了！怪不得两只小潜鸟长得这么快，这么健康，这么壮实呢。给双胞胎喂完食，加菲娅和潜鸟爸爸就连脑袋带身子整个儿沉到水下，不见了。接下来大约有两个小时，双胞胎奥莱尔连他们的影子都没见到，只能偶尔听见他们在唱歌，有时候唱的是《颤歌》，有时候是《笨歌》。

They were good children, and did not try to tag along or sulk because they were left behind. First they dabbled about and helped themselves, for dessert, to some plant growing under water, gulping down rather large mouthfuls of it. Then they grew drowsy; and what could have been pleasanter than going to sleep floating, with the whole cove for a cradle?

他俩都是乖孩子，不会因为爸爸妈妈不带他们就黏着追着，也不会噘嘴生气。他们先嬉了一会儿水，又找了一些长在水下的草，狼吞虎咽地当甜点吃了。后来，他们困了，漂在水上睡着了，整个小水湾就像是一个摇篮轻轻地摇着，世上难道还有比这更惬意的事情吗？

You could never guess how those youngsters got ready for their nap. Just like a grown–up! Each Olair rolled over on one side, till the white under–part of his body showed above water. Then he waved the exposed leg in the air, and tucked it away, with a quick flip, under the feathers of his flank. Thus one foot was left in the water, for the bird to paddle with gently while he slept, so that he would not be drifted away by the wind. But that day one of the tired water–babies went so sound asleep that he didn´t paddle enough, and the wind played a joke on him by shoving him along to the snaggy edge of the cove and bumping him against a log. That was a surprise, and he woke with a start and swam quickly back to the middle of the cove, where the other Olair was resting in the open water.

你永远都猜不出来这些小潜鸟是如何入睡的。其实和成年潜鸟一样！每只小潜鸟都是侧身翻到一边，露出白白的肚皮，然后，晃晃露在外面的那只腿，突然一缩把它藏到身侧羽毛的下面，于是，另一只脚就留在了水下，鸟儿睡觉的时候，就用它来轻轻地划水，这样他们就不会随风四处乱漂了。不过这天，一只水鸟宝宝太累了，睡得太沉，没有好好地划水，结果风和他开了个玩笑，把他吹到了暗桩丛生的小水湾边，撞上了一根大圆木。这一惊非同小可，他一下子就吓醒了，赶紧游回到小水湾的中央，那里，水域开阔，另一只奥莱尔正闭目养神呢。

While their children were napping, Gavia and Father Loon went to a party. On the way, they stopped for a bit of fishing by themselves. Gavia began by suddenly flapping around in a big circle, slapping the water with wing-tips and feet, and making much noise as she spattered the spray all about. Then she quickly poked her head under water, as if looking for fish. Father Loon, who had waited a little way off, dived a number of times, as if to see what Gavia had scared in his direction.

孩子们打盹儿小憩的时候，加菲娅和潜鸟爸爸赶去参加了一个聚会。路上，他们还停了片刻，自己捉了会儿鱼。加菲娅突然扑棱着翅膀转起了大圈，一边转，一边还用翅尖和脚掌拍打着水面，弄得水花四溅，发出噼里啪啦的响声。然后，她一头扎进水里，似乎在搜寻鱼儿。等在不远处的潜鸟爸爸，好几次潜入水中，似乎是想看看加菲娅把什么东西赶到他这边来了。

Then they both dove deep, and swam under water until they came near the four Neighbor Loons, who had left their two families of young dozing, and had also come out for a good time.

随后，他俩就潜到深处，在水下游了起来，一直游到那四个潜鸟邻居跟前，原来他们也是趁着孩子们打盹的时候，出来享受快乐时光的。

When Father Loon caught sight of his four neighbors, he sang the *Silly Song*, after which the six birds ran races on the water. They all started about the same time and went pell–mell in one direction, their feet and wings going as if they hardly knew whether to swim or fly, and ending by doing both at once. Then they would all stop, as suddenly as if one of them had given a signal, and turning, would dash in the opposite direction, racing to and fro again and again and again. Oh! It was a grand race, and there is no knowing how long

While their children were napping, Gavia and Father Loon went to a party.
当孩子们打盹儿小憩时，加菲娅和潜鸟爸爸赶去参加了一个聚会。

they would have kept it up, had not something startled them so that they all stopped and sang the *Tremble Song*, which sounds like strange laughter. They opened their mouths quite wide and, wagging the lower jaw up and down with every "ha". They sang "O, ha-ha-ha, ho! " so many times that it seemed as if they would never get through. And, indeed, how could they tell when the song was ended, for every verse was like the one before?

潜鸟爸爸看到他的四个邻居，就唱起了《笨歌》。一曲唱罢，六只鸟儿就开始在水面上比试赛跑。他们几乎是同时起跑的，然后匆匆忙忙地朝着同一个方向奔去，他们脚翅并用，好像都不知道是该游还是该飞了，结果只能是连游带飞了。突然，好像他们之中有谁发了信号似的，大家都停了下来，随即转身，又开始朝着来时的方向狂奔。就这样，他们来来回回，一次又一次地赛跑。哇！这真是一场盛大的比赛，要不是受到惊吓，让他们都停了下来，唱起了《颤歌》的话，根本无法知道他们会赛多久。他们的《颤歌》听起来像是古怪的笑声，每次唱“噢——”的时候，他们都把嘴张得大大地，下颚一上一下地晃动着；他们一遍又一遍地唱“噢，哈——哈哈，嗬！——”，唱了那么多遍，好像永远也唱不完似的。的确，他们怎么会知道这首歌何时结束呢？因为歌的每一节都是一样的！

Then all at once they stopped singing and began some flying stunts. A stiff breeze was blowing, and, facing this, they pattered along, working busily with wings and feet, until they could get up speed enough to leave the water and take to flight. Though it was rather a hard matter to get started, when they were once under way they flew wonderfully well, and the different pairs seemed to enjoy setting their wings and sailing close together around a large curve. They went so fast part of the time that, when they came down to the surface of the water again, they plunged along with a splash and ploughed a furrow in the water before they could come to a stop.

唱着唱着，他们突然一起停了下来，又开始表演飞行绝技。湖面的风吹得正急，他们迎着疾风，不停地挥动双翅、拍打双脚，啪嗒、啪嗒地往前冲，越冲越快，最后离开水面飞了起来。尽管起飞很难，但是，一旦离开水面，他们都飞得好极了！这几对潜鸟似乎都喜欢伸展双翅、亲密地依偎在一起滑翔，在天空中划出一道长长的曲线。有时候，他们飞得太快，结果回到水面的时候，由于用力过猛，会在水面上溅起一大片水花，冲出一道深深的浪沟，然后才能停下来。

Of course, by that time they were hungry enough for refreshments! So Gavia went off to one side and stirred the water up as if she were trying to scare fish toward the others, who waited quietly. Then they all dived, and what their black sharp-pointed bills found under water tasted good to those hungry birds.

当然，那个时候他们已经饥肠辘辘了，急需补充一些点心！于是加菲娅游到一边，开始搅动湖水，似乎是想把鱼都吓到其他几只潜鸟那边去，他们也正安静地等着呢。然后，所有的潜鸟一起潜入水里，他们已经饥饿难耐了，所以，无论他们那尖尖的黑色利嘴在水下找到了什么，对他们来说都是美味佳肴。

After that the loon party broke up, and each pair went to their own home cove, where they had left their young. It had been a pleasant way to spend the time sociably together; and loons like society very much, if they can select their own friends and have their parties in a wilderness lake. But gay and happy as they had been at their merrymaking, Gavia and her mate were not sorry to return to the two Olairs, who had long since wakened from their naps and were glad to see their handsome father and mother again.

吃完点心，潜鸟的聚会就结束了，双双回到各自位于小水湾的家中，那里还有他们的孩子呢。和大伙儿一起欢度时光真愉快呀！如果能选择自己的朋友，在湖泊的荒野世界里举办自己的聚会的话，潜鸟是非常喜欢与朋友们在一起的。但是，尽管刚才与朋友们狂欢时他们开心极了，但是，回到两个奥莱尔宝宝身边，加菲娅夫妇一点儿也不遗憾。两个宝宝已经睡醒好久了，又见到了漂亮的爸爸妈妈，别提多高兴了！

By the time the two Olairs were full grown, Gavia had molted many of her prettiest feathers and was looking rather odd, as she had on part of her summer suit and part of her winter one. Father Loon had much the same appearance; for, of course, birds that live in the water cannot shed their feathers as many at a time as Corbie could, but must change their feather-wear gradually, so that they may always have enough on to keep their bodies dry. And summer and winter, you may be sure that a loon takes good care of his clothes, oiling them well to keep them waterproof.

等到两个奥莱尔宝宝完全长大时，加菲娅身上那漂亮的羽毛很多都已经换掉了，看上去有点奇怪，因为她身上穿的一半是夏装，一半是冬装。潜鸟爸爸的样子也差不了多少。当然了，住在水上的鸟儿无法像乌鸦考比那样一次能掉很多羽毛，他们必须慢慢地换掉身上的羽衣，这样身上才会有足够的羽毛让身体保持干燥。放心吧，无论冬夏，潜鸟都会悉心呵护身上的衣服，精心地为它上油，以保持它的防水功能。

Fall grew into winter, and the nest where Gavia had brooded the spring before now held a mound of snow in its lap. The stranded log against which the little Olair had been bumped while he was napping, months ago, was glazed over with a sparkling crust. The water where Gavia and Father Loon had fished for their children, and had played games and run races with Neighbor Loons, was sealed tight with a heavy cover of ice.

秋去冬来，春天加菲娅孵蛋的那个鸟巢现在盛着满满一堆雪；好几个月前小奥莱尔打盹时撞上的那根搁浅的木头，上面结着一层晶莹闪亮的雪壳；加菲娅和潜鸟爸爸曾经为孩子们捕鱼、与邻居们嬉戏比赛的那片水域被一层厚厚的冰封了个严严实实。

And it may be, if you should sail the seas this winter, that you will see the two Olairs far, far out upon the water. What made them leave the pleasures of Immer Lake just when they did, I cannot explain. I do not understand it well enough. I never felt quite sure why Peter Piper left the shore where the cardinal flowers glowed, for far Brazil. All I can tell you about it is that a feeling came over the loons that is called a migration instinct, and, almost before they knew what was happening to them, they were laughing weirdly through the ocean storms.

如果这个冬天你去海里航行，也许会在远远的海面上看到那两个奥莱尔呢。是什么让他们放弃了爱默湖的快乐生活，我无法解释，因为我了解的也不够。我从来也不是很清楚，为什么矶鹞彼得要离开盛开着红花半边莲的海岸飞往遥远的巴西。我只知道，潜鸟突然就有了一种感觉，这种感觉叫作“迁徙直觉”。他们几乎还没明白是怎么回事的时候，就已经迎着海上的狂风暴雨奇怪地大笑起来了。

If you see them, you will know that they are strange birds whose ancestors reach back and back through the ages, maybe a million years. You will think—as who would not?—that a loon is a wonderful gift that Nature has brought down through all the centuries; a living relic of a time of which we know very little except from fossils men find and guess about.

如果你看到他们，你就会知道他们就是那种奇怪的鸟，其祖先可以跨越无数个时代，追溯到很远、很远，也许是一百万年以前。你会想——谁不会这样想呢？——潜鸟是大自然历经千百万年送给我们的奇妙礼物；是远古时期留给我们的活纪念物，关于那个时期，除了人们从化石上发现的东西和猜测的情况之外，其他我们真的是所知甚少。

It is small wonder their songs sound strange to our ears, for their voices have echoed through a world too old for us to know. It makes us a bit timid to think about all this, as it does the minister of Immer Lake, who sits before his door through many a summer twilight, playing on his violin until the loons answer him with their *Tremble Song*:—

"O, ha-ha-ha, ho! O, ha-ha-ha, ho! "

所以，他们的歌声我们听起来怪怪的，也就不足为奇了，因为他们的声音是穿越了一个古老世界的回声，那是一个古老得我们无法了解的世界。想到这些，我们就会心生敬畏，就像住在爱默湖畔的那位牧师一样，夏日里，许许多多个黎明薄暮中，他坐在门前，拉着他的小提琴，等着潜鸟唱起《颤歌》来应和他的琴声：

“噢，哈——哈哈，嗬！哈——哈哈，嗬！”

5

Eve and Petro
燕子伊娃和佩特罗

If swallows studied history, 1920 would have been an important date for Eve and Petro. It was the one hundredth anniversary of the year when a man named Long visited cliff swallows among the Rocky Mountains.

如果燕子们也研究历史的话，那么 1920 年对伊娃和佩特罗来说就是一个重要的年代，因为，曾经有一位名叫朗的人探访过落基山脉的崖燕，而 1920 年距那次探访刚好一百周年。

The century between 1820 and 1920 had given what we call civilization a chance to make many changes in the wild world of birds. During that time lifeless hummingbirds had been made to perch upon the hats of fashionable women; herring gulls had been robbed of their eggs and killed for their feathers; shooting movements had been organized to kill crows with shotgun or rifle, in order that more gunpowder might be sold; the people of Alaska had been permitted to kill more than eight thousand eagles in the last great breeding-place left to our National Emblem; uncounted millions of Passenger Pigeons had been slaughtered, and these wonderful birds done away with forever; and the methods by which egrets had been murdered were too horrible to write about in books for children to read.

1820 年到 1920 年的这一百年里，我们所谓的文明终于有机会得以多方改变鸟类赖以生存的野生世界。那个时期，失去生命的蜂鸟被安放栖息在时尚女士的帽子上；人们不仅攫取银鸥产下的蛋，还杀死他们拔走他们身上的羽毛；人们组织了一次又一次的打猎运动，用散弹枪和步枪射杀乌鸦，为的是卖掉更多的火药；鹰是美国的象征，可是，阿拉斯加的居民却获准在鹰的最后一块主要繁衍地上将其杀害，数量高达八千只以上；成百上千万的候鸽遭到屠杀，数也数不清，最终这种奇妙的鸟儿永远地绝迹了；人类谋杀白鹭的手段更加恐怖，不适合孩子们阅读，这里就不写了。

But however shamefully civilization had treated, and had brought up children to treat, these and many other of their fellow creatures of the world, who had a right to the life that had been given them as surely as it had been given to men, the years since 1820 had been happy ones for the ancestors of Eve and Petro.

这些鸟，以及生活在这个世界上的其他许许多多鸟类，他们都有享受生命的权力，就如同人类有享受生命的权力一样。尽管人类的文明世界如此无耻地对待他们，并让自己的后代也步其后尘，但是，1820 年以后的那些年，对于燕子伊娃和燕子佩特罗的祖先来说还是幸福的。

Eve and Petro, themselves, were happy as any two swallows need be that spring of 1920, when they started forth to seek a cliff, just as their ancestors had done for the hundred years or so since man began to notice their habits, and no man knows for how many hundreds of years before that.

1920 年的春天，当伊娃和佩特罗追随祖先们的习惯，动身去寻找悬崖峭壁的时候，他们和所有出双入对的燕子一样，开心极了。人类注意到燕子的这种习惯才有一个世纪左右，在这之前这个习惯已经沿袭了多少个世纪，却无人知晓。

Of course they found it as all cliff swallows must, for cliff-hunting is a part of their springtime work. It was very high and very straight. Its wall was of boards, and the gray shingled roof jutted out overhead just as if inviting Eve and Petro to its shelter.

当然了，和所有崖燕一样，伊娃和佩特罗找到了自己的峭壁，因为这是春忙的一部分。这个地方又高又陡，壁面是木板搭成的，盖着木瓦的灰色的屋顶向外探伸着，像是在邀请他们进去躲避风雨。

It was a good cliff, and mankind had been so busy building the same sort all across the country for the past hundred years that there was no lack of them anywhere, and swallows could now choose the ones that pleased them best. Yes, civilization had been kind to them and had made more cliffs than Nature had built for them; though perhaps it was Mother Nature, herself, who taught the birds that these structures men called barns and used inside for hay or cattle were, after all, only cliffs outside, and that people were harmless creatures who would not hurt the swallow kind.

这个“峭壁”还真不错，过去的一百年里，人类忙着到处建造这样的“峭壁”，所以，现在这样的“峭壁”随处可见，燕子们可以随心所欲精挑细选。的确，文明世界对燕子还不错，他们为燕子建造的“峭壁”，比大自然母亲为燕子提供的峭壁可多多了。不过话说回来，也许就是大自然母亲自己告诉燕子，说这些被人类叫作谷仓，里面用来存干草或者关牲畜的建筑，外面的墙其实就是“峭壁”，也许大自然母亲还告诉燕子，说人类这种动物对燕子没有恶意，不会伤害他们。

However all that may be, it is quite certain that Eve and Petro squeaked pleasantly for joy when they chose their building site, undisturbed by the ladder that was soon put near, and unafraid of the people who climbed up to watch them at their work. They were too happily busy to worry, and besides, there is a tradition that men folk and swallow folk are friendly, each to the other.

反正不管是怎么回事,有一点可以肯定,那就是伊娃和佩特罗选定筑巢地点后，开心地叽叽喳喳叫了起来。没过多久,就有一个梯子架到了跟前,可是,他们一点儿也不在意，接着，有人爬上梯子来看他们筑巢,他们也一点儿都不害怕。太忙了,太高兴了,已经顾不上担心了;再说了,人类和燕子还有相互友爱的传统呢。

How old this tradition is, we do not know; but we do know that swallows of one kind and another were welcomed in the Old World in the old days to heathen temples before there were Christian churches, and that today in the New World they play in and out of the dark arches in the great churches of far Brazil and flash across the gilding of the very tabernacle, reminding us of the passage in the Psalms where it is written that the swallow hath found a nest for herself, where she may lay her young—even thine altars, O Lord of Hosts!

这个传统的历史有多悠久,我们不得而知;不过我们知道,在昔日的旧大陆(注:指在哥伦布发现新大陆之前，欧洲认识的世界，包括欧洲、亚洲和非洲），还没有基督教教堂的时候，无论是哪种燕子只要飞进庙宇里，都会受到欢迎；我们还知道，如今在新大陆（注：美洲大陆及邻近的群岛），在遥远的巴西，在那些宏伟的教堂里，燕子们在昏暗的拱廊上进进出出、嬉戏玩耍，在镀金的神龛上轻盈飞过，这情形让我们不由地想起《圣经》里的诗句：万军之耶和华，在你的祭坛里，燕子为自己找到了抱雏之窝。

So it is not strange that far and wide over the world people have the idea that swallows bring luck to the house. I think so myself, don´t you?—that it is very good fortune, indeed, to have these birds of friendly and confiding ways beneath our shelter.

所以，全世界所有地方的人都认为燕子的到来会带来好运，这就不足为怪了。我也这样认为，你呢？这些对我们充满了友爱和信任的鸟儿能住在我们的屋檐下，的确是我们的好运气呀！

Of course the ancestors of cliff swallows had not known the walls and roofs of man so long as other kinds of swallows; but the associations of one short century had been pleasant enough to call forth many cheerful squeakings of joy, just like those of Eve and Petro that pleasant day in June when they started their nest under the roof near the top of the ladder.

当然了，和其他燕子相比，崖燕的祖先们认识人类的墙壁和屋顶的时间并不长；但是，短短一个世纪的接触非常愉快，所以，一看见这些墙壁和屋顶，他们就禁不住叽叽喳喳地欢叫起来。伊娃和佩特罗就是这样，六月的那一天，当他们准备在那个梯子附近的屋顶上筑巢的时候，就不由地发出了叽叽喳喳的欢叫。

To be sure, they made no use of that ladder, even though they were masons and had their hods of plaster to carry way up near the top of their cliff. No, they needed no firmer ladder than the air, and their long wings were strong enough to climb it with.

虽然他们是泥瓦匠，需要把一桶一桶的泥浆搬到高高的壁顶上去，但是，可以肯定，他们根本就没用那个梯子。是的，他们不需要那个梯子，因为对他们来说没有比高空更结实的梯子了；而且，他们长长的翅膀很发达，爬这个梯子没问题。

They lost little time in beginning, each coming with his first hod of plaster. How? Balanced on their heads as some people carry burdens? No. On their backs, then? No. In their claws? Oh, no, their feet were far too feeble for bearing loads. Do you remember what Corbie used for a berry-pail when he went out to pick fruit? Why, of course! the hod of the swallow mason is none other than his mouth, and it holds as much as half a thimbleful.

来的时候,他们俩各带着自己的第一桶泥浆,所以,一点儿工夫都没耽误,就开始动工了。怎么带的呢?像有些人那样顶在脑袋上?不是的。那就是背在背上?也不是。抓在爪子里?噢,更不是了,他们的爪子太小了,根本就没劲拿东西。你还记得乌鸦考比去摘果子的时候用什么当果桶的吗?没错,当然了!燕子的泥浆桶不是别的,就是他的嘴巴,他的嘴巴能盛比一丁点儿还少一半的泥浆呢。

First, Eve had to mark the place where the curved edge of the nest would be; and how could she mark it without any chalk, and how could she make a curve without any compasses? Well, she clung to the straight wall with her little feet, which she kept nearly in one place, and, swinging her body about, hitch by hitch, she struck out her curve with her beak and marked it with little dabs of plaster. Then she and Petro could tell where to build and, taking turns, first one and then the other, they began to lay the wall of their home.

首先,伊娃得标出巢口弧形外沿的位置。可是,没有粉笔,怎么标位置呢?没有圆规,又怎么画曲线呢?嗨,没关系,伊娃双脚紧紧地钩住笔直的墙面,固定住,然后一点一点移动身体,用嘴巴啄出需要的曲线,再涂上泥点标出来,这样她和佩特罗就知道该在哪里筑巢了。接下来,他们就开始轮流为自己的家砌墙了,你砌完,我再砌。

It was slow work, for it must be thick and strong, and the place where they gathered the plaster was not handy by, and it took a great great many trips, their hods being so small.

这可是个慢功夫活,因为巢壁必须又厚又结实;而且,采集泥浆的地方不在跟前,一点儿都不方便,再说,他俩的"泥浆桶"那么小,得一趟又一趟地出去衔泥。

At first, while the nest was shallow, only one could work at a time; and if Petro came back with his plaster before Eve had patted the last of hers into place, she would squeak at him in a fidgety though not fretful voice, as if saying, "Now, don´t get in my way and bother me, dear." So he would have to fly about while he waited for her to go. The minute she was ready to be off, he would be slipping into her place; and this time she would give him a cosy little squeak of welcome, and he would reply, with his mouth full of plaster, in a quick and friendly way, as if he meant, "I´ll build while you fetch more plaster, and we´d both better hurry, don´t you think?"

刚开始巢浅的时候，一次只能有一只燕子砌墙；要是伊娃还没有把她的泥浆抹利索，佩特罗就带着泥浆回来了，她就会冲他叽叽喳喳地叫，听上去并不烦躁，但却有几分不安，似乎在说“喂，亲爱的，别来碍我的事，别来烦我”。于是，佩特罗就只好在附近飞来飞去，等着她完工。伊娃刚要离开，佩特罗就会溜进去把她挤到一边。这次，她也冲着他低低地叽喳了一声，不过是在表示欢迎，听上去温和多了。虽然佩特罗的嘴巴里衔满了泥，但是，他也会立刻亲切地回应，好像在说：“你去取泥的时候，我会砌墙的。咱俩最好都快点，你说呢？”

After worrying a bit about the best place to dump his hodful, he went to work. He opened his beak and, in the most matter–of–fact way, pushed out his lump of plaster with his tongue, on top of the nest wall. Then he braced his body firmly in the nest and began to use his trowel, which was his upper beak, pushing the fresh lump all smooth on the inside of the nest.

佩特罗略微踌躇了一下，才找到了一个最佳位置把自己的那“桶”泥浆倒了出来。只见他张开嘴，泰然自若地伸出舌头，把嘴里的泥团吐在了巢壁的最上面。接着，他在巢里面稳稳地撑住身子，用他的小铲子——也就是他的上喙——把刚吐上去的泥团平展展地抹在了巢的内壁上。

Have you ever seen a dog poke with the top of his nose, until he got the dirt heaped over a bone which he had buried? Well, that´s much the way Petro bunted his plaster smooth—rooted it into place with the top of his closed beak. He got his face dirty doing it, too, even the pretty pale feather crescent moon on his forehead. But that didn´t matter. Trowels, if they do useful work, have to get dirty doing it, and Petro didn´t stop because of that. If he had, his nest would have been as rough on the inside as it was outside, where a humpy little lump showed for each mouthful of plaster.

你见过狗为了藏骨头用鼻头拱土的情形吗？他拱呀拱，一直拱出一个土堆才罢休。其实，佩特罗抹泥用的就是这种方法：他闭起嘴巴，把泥拱起来抹到墙上。抹泥的时候，他也会把脸弄脏，就连额头上那块漂亮的新月形灰白色羽毛也被弄得脏乎乎的。不过，这无所谓，因为，要让“小铲子”有所作为，那儿就干净不了，佩特罗才不会因为这个停工呢。否则，巢的内壁早就和外壁一样，一口泥一个小疙瘩，别提多粗糙了。

Although Eve and Petro did not fly off to the plaster pit together, they did not go alone, for there was a whole colony of swallows building under the eaves of that same barn; and while some of them stayed and plastered, the rest flew forth for a fresh supply.

虽然伊娃和佩特罗没有一同飞往泥坑，但是，他们并不孤单，因为在那个谷仓的仓檐下，还有一大群燕子也在筑巢；一些燕子留下来抹泥的时候，另一些燕子就飞去衔新泥。

They knew the place, every one of them; and swiftly over the meadow and over the marsh they flew, until they came to a pasture. There, near a spring where the cows had trampled the ground until it was oozy and the water stood in tiny pools in their hoof prints, the swallows stopped. They put down their beaks into the mud and gathered it in their mouths; and all the time they held their wings quivering up over their beautiful blue backs, like a flock of butterflies just alighting with their wings atremble.

他们知道那个地方，大家都知道；他们翩翩而飞，越过草地，越过沼泽，来到一片牧草地上，落在一眼泉水旁。这里的地已经被母牛踩得软乎乎、湿答答的，牛蹄印里积满了水，变成了一摊摊的小水洼。他们把嘴巴伸进泥浆，衔一些到嘴里；这期间，他们的翅膀一直高高地举在漂亮的蓝色脊背上抖个不停，像一群飘然而落的蝴蝶，不停地忽闪着翅膀。

So their plaster pit was just a mud-puddle. Yes, that is all; only it had to be a particularly sticky kind of mud, which is called clay; for the walls of their homes were a sort of brick something like that the people made in Egypt years and years ago. And do you remember how the story goes that the folk in Pharaoh´s day gathered straws to mix with the clay, so that their bricks would be stronger? Well, Eve and Petro didn´t know that story, but they gathered fibres of slender roots and dead grass stems with their clay, which doubtless did their brick plaster no harm.

原来,他们的泥浆坑只不过是一个泥洼呀。是的,如此而已;只不过那泥得特别黏,得是被称作黏土的那种泥才行;因为他们家的墙就是一种砖,有点像是许许多多年以前埃及人制作的那种砖。你记得故事里是怎么说的吗？为了让砖更结实,法老时期的人们搜集稻草掺在黏土里。当然了，伊娃和佩特罗可不知道这个故事，可是,他们却把细须根和枯草茎的纤维搜集起来,掺在黏土里。毫无疑问,这种做法不会破坏他们的砖泥。

Men brick–makers nowadays bake their bricks in ovens called kilns, which are heated with fire. Eve and Petro let their brick bake, too, and the fire they used was the same one the Egyptians used in the days of Pharaoh—a fire that had never in all that time gone out, but had glowed steadily century after century, baking many bricks for folk and birds. Of course you know what fire that is, for you see it yourself every day

At Work in the Plaster Pit.
在泥浆坑里衔泥。

that the sun shines.

如今，人类的制砖人是用火给炉子加热来烧砖的，烧砖的火炉叫砖窑。伊娃和佩特罗的砖也要烧，他们用的火就是法老时期埃及人用的那个火——那个不仅从来没有熄灭过，而且在一个世纪又一个世纪的时间里，一直稳稳当当地发着光、发着热，为人们和鸟儿们烧制墙砖。当然了，你知道那是什么火，因为每天，只要阳光明媚，你都会看见它。

Every now and again Eve and Petro and all the rest of the swallow colony left off their brick-building and went on a hunting trip. They hunted high in the air and they hunted low over the meadow. They hunted afar off along the stream and they hunted near by in the barnyard. And all the game they caught they captured on the wing, and they ate it fresh at a gulp without pausing in their flight. As they sailed and swirled, they were good to watch, for a swallow´s strong long wings bear him right gracefully.

时不时地，伊娃、佩特罗和整个燕子大军都会停下砌砖的活儿，踏上捕食之旅。他们在高高的空中捕食，也在低低的草地上捕食；他们沿着远处的小溪捕食，也在附近谷仓旁的空地上捕食。所有的猎物都是边飞边捕的,而且,一边飞着,一边就趁着新鲜一大口吞进了肚里。他们在空中飞翔盘旋的时候,非常好看,因为,燕子的翅膀长长的,很发达,飞起来十分优雅。

Why did they stop for the hunting flight? Perhaps they were hungry. Perhaps their mouths were tired of being hods for clay they could not eat. Perhaps the fresh plaster on the walls of their homes needed time to dry a bit before more was added.

他们为什么要停工飞去觅食呢?也许是他们饿了;也许是他们的嘴厌倦了,不想再衔那些不能吃的黏土了;也许是刚抹到他们家墙壁上的泥巴需要一段时间才能干,所以暂时不需要去衔泥。

Be that as it may, they made the minutes count even while they rested from their building work. For they used this time getting their meals; and whenever they were doing that, they were working for the owner of the barn, paying their rent for the house-lot on the wall by catching grass insects over the meadow, and mosquitoes and horseflies and house–flies by the hundreds, and many another pest, too.

总之，不管怎样，就算是停工休息的时候，他们也是分秒必争，因为他们总是利用这点儿时间去吃饭，而且，每次吃饭，他们都是在为谷仓的主人工作，因为他们会捕捉草地上的昆虫，还有成百成百的蚊子、马蝇、家蝇和许许多多其他害虫，以此来偿付筑巢占用墙壁的租金。

Ah, yes, there may be some reason for the belief that swallows bring good luck to men. I once heard of a farmer who said he didn´t dare disturb these birds because of a superstition that, if he did, his cows wouldn´t give so much milk. Well, maybe they wouldn´t if all the flies a colony of swallows could catch were alive to pester his herd; for the happier and more comfortable these animals are, the healthier they are and the more milk they give.

啊，是的，认为燕子会给人类带来好运也许有点道理。有一次，我听说有一个农夫说他不敢打扰燕子，因为他很迷信，他认为如果他打扰了燕子，那么他家的奶牛就不会产那么多奶了。的确，如果燕群捕食的那些蝇

The Hunting Flight
捕食之旅

虫都活着，不停骚扰他家的奶牛的话，它们也许真的就产不了那么多奶了，因为奶牛越快乐、越舒服，就越健康，产的牛奶也就越多。

The hunting flights of Eve and Petro and their comrades lasted about fifteen minutes each time they took a recess from their building.

伊娃、佩特罗和伙伴们停工期间的捕食之旅，大约会持续十五分钟左右。

After two days the nest was big enough, so that there was room for both swallows to build at once; and after that, Petro didn´t have to fly around with his mouth full of plaster waiting for Eve to go if he chanced to come before she was through. They always chatted a bit and then went on with their work, placing their plaster carefully and bunting it smooth on the inside, modeling with clay a house as well suited to their needs as is the concrete mansion a human architect makes suited to the needs of man.

燕巢越建越大，两天之后，两只燕子都可以容身其中一起施工了。从此，如果佩特罗回来的时候，碰上伊娃还没有完工，他再也不用衔着满嘴的泥，在周围飞来飞去等她离开了。他俩总是先聊几句，然后再接着干活。先小心翼翼地把嘴里的泥放到位，接着平平展展地抹在巢的内壁上，他们用黏土砌成的巢和人类用水泥建成的大厦一样功能齐全、适合居住。

And if you think it is a simple matter to make a nest of clay, just go to the wisest architect you know and ask him these questions. How many hodfuls of clay, each holding as much as half a thimble, would it take to build the wall of a room just the right shape for a swallow to sit in while she brooded her eggs? How large would it have to be inside, to hold four or five young swallows grown big enough for their first flight? How thick would the walls have to be to make it strong enough? What sort of curve would be best for its support against a perfectly straight wall? How much space would have to be allowed for lining the room, to make it warm and comfortable? How can the clay be handled so that the drying sun and wind will not crack the walls? What is the test for telling whether the clay is sticky enough to hold together? How much of the nest must be stuck to the cliff so that the weight of it will not make it fall?

如果你认为筑一个泥巢很容易的话，那就找你知道的最高明的建筑师问问吧，看看他能否回答这些问题：假如每只“泥浆桶”只能盛一丁点儿一半的泥，那么一共需要多少“桶”泥才能砌完一个形状正好适合燕子孵蛋卧的巢？巢的里面需要有多大空间才能容得下四五只小燕子发育成长？巢壁要多厚才能结实耐用？要支撑在完全笔直的墙上，巢壁的最佳弧度是多少？为了让巢又暖和又舒适，需要预留多少空间做铺垫？黏土需要怎样处理，砌成的墙壁才能经得起风吹日晒而不致裂缝？怎么才能测出泥的黏度达到了要求？巢与“峭壁”的粘连面积要达到多少才能挂住巢身，不让它掉下来？

If the architect can answer all those questions, ask him one more: ask him if he could make such a nest with the same materials the birds used, and with no more tools?

如果建筑师能回答所有这些问题，那就再问问他，看他能不能用燕子用的材料，不用别的工具，建造一个这样的鸟巢？

Well, Eve and Petro could and did. It was big enough and strong enough and shaped just right; and when it was nearly done and nearly ready for the soft warm lining, That Boy climbed the ladder and knocked it down with his hand.

伊娃和佩特罗就能,而且做到了。他们筑的巢够大、够结实,形状也恰到好处。可是,鸟巢即将竣工,眼看着就可以往里面铺松软暖和的垫子时,“那个男孩”爬上梯子,用手把它敲了下来。

There it lay, Eve and Petro´s wonderfully modeled nest of clay, broken to bits on the ground and spoiled, oh, quite spoiled. There is a saying that it brings bad luck to do harm to a swallow. What bad luck, then, had the hand of That Boy brought to the world that day?

伊娃和佩特罗精心筑造的泥巢，就这样躺在了地上,摔成了碎片,完了,唉,全完了。俗话说伤害燕子会倒霉的。那么,那一天,“那个男孩”的手给这个世界带来的是什么倒霉的事情呢?

They always chatted a bit and then went on with their work, placing their plaster carefully.
他们总是先聊几句，然后接着干活，小心翼翼地把嘴里的泥放到位。

Bad luck it brought to Eve and Petro, who had toiled patiently and unafraid beside the ladder–top, with faith in those who climbed quietly to watch the little feathered masons at their work. But now the walls of their home were broken and crumbled, and their faith was broken and crumbled, too. In dismay they cried out when they saw what was happening, and in dismay their swallow comrades cried out with them. Fear and disappointment entered their quick hearts, which had been beating in confidence and hope. People who climbed ladders were not beings to trust, after all, but frightful and destroying creatures. This had the hand of That Boy brought to Eve and Petro, who looked at the empty place where their nest had been, and went away.

这只手给伊娃和佩特罗带来了厄运。这两个满身羽毛的小泥瓦匠一直不辞辛劳，耐心地在梯顶附近忙碌着，一点儿也不害怕那些悄悄地爬上来看他们筑巢的人，因为他们对那些人充满了信任。可是现在，他们的家被摔烂了，成了碎片，他们的信任也被摔烂了，成了碎片。看到发生的一切，他俩惊恐万状，不由地大叫起来，他们的伙伴们也惊恐万状，和他们一起大叫起来。恐惧和沮丧慑住了他们敏感的心，那曾经跳动着信任和希望的心。原来，那些爬上梯子的人是不可以信赖的，他们很可怕，会破坏一切。这就是那个男孩的手带给伊娃和佩特罗的厄运。看着巢去壁空的地方，他俩飞走了。

Bad luck it brought to an artist who drew pictures of birds; and when he knew what had happened, a sudden light flamed in his eyes. The name of this light is anger—the kind that comes when harm has been ruthlessly done to the weak and helpless. For the artist had climbed the ladder many a time, and had laid his quiet hand upon the lower curve of the nest while Eve and Petro went on with their building at the upper edge. And he had seen the colors of their feathers and the shape of the pale crescent on their foreheads—the mark a man named Say had noticed many years before, when he named this swallow in Latin, *lunifrons*, because *luna* means moon and *frons* means front. And he had hoped to climb the ladder many a time again, and when there should be young in the nest, to see how they looked and watch what they did, so that he could draw pictures of the children of Eve and Petro.

这只手给一位画鸟儿的画家带来了厄运。当他得知发生的一切时，双眼一下子燃起了熊熊的烈焰，这烈焰的名字就叫怒火——这是弱小无助者被无情伤害时就会燃烧的一种烈焰，因为这位画家也曾多次爬上梯子，轻轻用手触摸巢底的曲线，每次，伊娃和佩特罗都旁若无人地继续在巢的上沿忙碌着。他见过他俩身上羽毛的颜色，还有他俩前额上那浅浅的像新月一样的印记——这个印记是很多年以前一个叫赛的人给这种崖燕命名的时候注意到的，他用拉丁语给这种崖燕起了个名字，叫"lunifrons"，因为"luna"是月亮的意思，"frons"的意思是"前面"。他当初还希望他能再多爬上去几次，等到窝里孵出小燕子时，看看他们长得什么样，观察一下他们都干什么，这样他就能画伊娃和佩特罗的宝宝们了。

Bad luck it brought to a writer of bird stories; and when she knew what had happened, something like an ache in her throat seemed to choke her, something that is called anger—the kind that comes when harm is done to little folk we love. For she had climbed the ladder many a time, and had rested her head against the top while she watched Eve and Petro push the pellets of mud from their mouths with their tongues and bunt the wall of their clay nest smooth on the inside with the top of their closed beaks, not stopping even though they brushed their pretty chestnut-colored cheeks against the sticky mud, or got specks on the feathers of their dainty foreheads that bore a mark shaped like a pale new moon. And she had hoped to climb the ladder many a time again, and watch Eve and Petro feed their children when the nest was done and lined and the eggs were laid and hatched; for this nest could be looked into, as the top was left open because the barn roof sheltered it and it needed no other cover.

这只手给一位专门写鸟故事的作家带来了厄运。当她得知发生的一切时，一股气一下子冲上了喉头，几乎让她窒息。这股气的名字就叫愤怒——这是我们心爱的孩子受到伤害时就会冲出的一股气。因为女作家曾经多次爬到梯子上，下巴颏搭在梯子的顶端，看着伊娃和佩特罗伸出舌头，从嘴里吐出一颗颗泥粒，然后，闭上嘴巴，用嘴巴的上侧把泥巢的内壁推抹平整。他们一刻也不停地忙碌着，就算是漂亮的栗色脸蛋擦到了黏乎乎的泥巴上，或者就算是泥点溅到了印着浅浅的像新月一样印迹的精致的前额上，弄脏了那里的羽毛，他们也不停工。她当初还希望能再多爬上去几次，等他俩筑完巢铺好垫子、生出蛋孵出小燕子的时候，看着他们给小宝宝喂食呢！要知道，他们筑的这个巢上面敞开着，无遮无拦，谁都可以窥视，因为有谷仓顶为他们的巢遮风避雨，就不需要其他遮盖了。

Now Eve and Petro were gone, and no more sketches could be made near enough to show how little cliff swallows looked in their nest. And nothing more could be written about such affairs of these two birds as could only be learned close to them. Nor, indeed, was there any way to learn those things from the rest of the colony; for it so chanced that Eve and Petro were the only pair who had built where a ladder could be placed. So bad luck had come not only to Eve and Petro, but to the story of their lives.

现在伊娃和佩特罗飞走了，无法再画近距离的素描了，也就无法向人们展示泥巢里小崖燕的模样了；关于这两只燕子的有些事情也无法再写了，因为这些事情只有靠近他们才能了解，而且，这些事情也无法从其他的燕群那里知晓，因为，很不凑巧，只有伊娃和佩特罗这对崖燕把巢筑在了一个可以放梯子的地方。所以遭厄运的不仅是伊娃和佩特罗，还有他俩的故事。

But, most of all, the breaking of their nest brought bad luck to That Boy, himself. For as he stood at the top of the ladder, he might have curved the hollow of his hand gently upon the rounded outside of the nest and, waiting quietly, have watched the building birds. He might have seen Eve come flitting home with her tiny load of clay, poking it out of her mouth with her tongue and bunting it smooth in her own cunning way. He might have laid his head against the ladder and heard their cosy voices as they squeaked pleasantly together over the home–building. He might have looked at the colors of their feathers, and seen where they were glossy black with a greenish sheen, where rich purply chestnut, and where grayish white. He might have looked well at the pale feather moon on their foreheads, which the man named Say had noticed one hundred years before. He might, oh, he might have become one of the brotherhood of men, whom swallows of one kind or another have trusted since the far–off years of Bible times when they built at the altars of the Lord of Hosts.

不过，最糟糕的是，把他俩的泥巢弄碎让那个男孩自己也倒了霉。因为当他站在梯顶的时候，也许他本可以用手心轻轻地捧住圆乎乎的巢外壁，悄悄地待着，观看这对筑巢的鸟儿；也许他本可以看着伊娃衔着她的那一小份泥，轻快地飞回家，熟练地伸出舌头从嘴里吐出来，再把它抹平；也许他本可以把脑袋搭在梯子上，听他们一边筑巢，一边愉快地叽叽喳喳，那声音温馨极了。也许他本可以观察一下他们身上羽毛的颜色，看看哪里的羽毛黑油油的，还泛着绿光，看看哪里的羽毛是深栗色的，还带着一点浅浅的紫色，还可以看看哪里的羽毛是灰白色的。也许他本可以把一百年前那个叫赛的人发现的他们前额上的那块浅浅的形状像月亮一样的羽毛看个清清楚楚；也许他，哦，也许他本可以像其他情同手足的人类一样，得到燕子的信赖，因为从遥远的《圣经》时代开始，从燕子在万军之主耶和华的圣坛旁筑巢开始，人类就得到了所有燕类的信赖。

All this good luck he held, That Boy, in the hollow of his hand, and he threw it away when he struck the nest; and it fell, crumbled, with the broken bits of clay.

这么多的好运气都捧在那个男孩的手心，可是，就在他打破燕巢那一刻，他也就把这些好运给扔掉了；随着跌碎的黏土片，这些好运也掉到了地上，摔成了碎片。

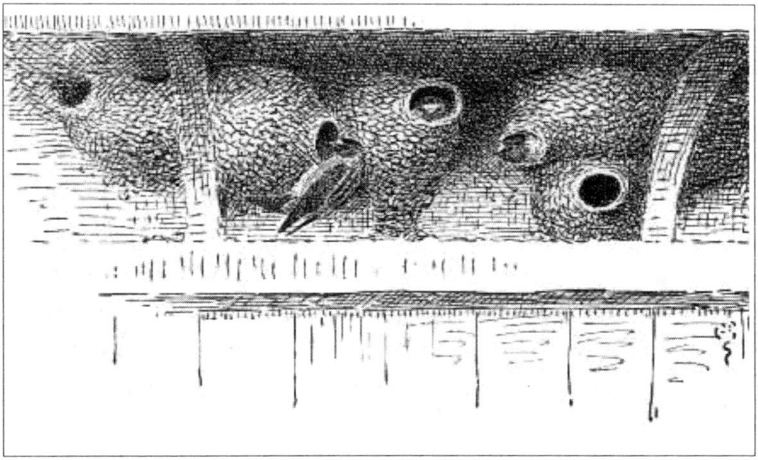

Quaint Clay Pottery
模 样 古 怪 的 土 陶 器 似 的 燕 巢

As for Eve and Petro, if fear and disappointment had driven trust from their hearts, they still had courage and patience and industry. They sought another and a different sort of cliff, and found one made of red brick and white stone. Near the very high top of this a large colony of swallows were building; and, because there was no closely protecting roof, these swallows were making the round part of their nest closed over at the top with a winding hallway to an outer doorway. They looked, indeed, like a row of quaint clay pottery, shaped like crook-necked gourds. For such were the nests these swallows built one hundred years ago on the wild rock cliffs, if they chose their house-lots where there was no overhanging shelter; and such are the nests they still build when there seems to be need of them.

至于伊娃和佩特罗，尽管恐惧和沮丧赶走了他们内心对人类的信赖，但是他们还有勇气、有耐心、有勤奋。他们又去寻找另一种不同的峭壁，而且找到了，那是一块红砖白石做成的峭壁，高高的壁顶附近，一大群崖燕正筑巢呢。因为巢的上方没有密实的屋顶做保护，所以这些燕子做的巢中间圆鼓鼓的，然后往上慢慢地收拢，最后收成了一个迂回曲折的通道，直通巢口，看上去就像是一排精致的泥质陶器。的确很像，它们的形状和歪脖子葫芦差不多。一百年前，这些燕子在荒郊野外的岩壁上筑巢时，如果选择筑巢的地方没有现成的“屋檐”，他们筑的就是这种巢。时至今日，如果需要，他们仍旧会筑这种巢。

They were too far from the pleasant pasture to dig their clay out of the footprints of cows; but there was a track where the automobiles slushed through sticky mud, and they swirled down there and filled their little hods when the road was clear.

他们离那个开心牧场太远了，没法从奶牛的蹄印里啄取黏土。不过附近有一条泥泞的小路，汽车驶过的时候就会泥水四溅，于是等到路面清静的时候，他们就会盘旋而下，把他们的“小桶”装满。

Eve and Petro found a nook even higher up than the others, where a crook-necked jug of a nest did not seem to fit. When they had built their wall as high as need be, they closed it over with a little rounded dome, and at the side they left two doorways open, one facing the southwest and one facing the southeast.

And some days after this was done, had you gone to the foot of their cliff and used a pair of field-glasses, you might have seen Eve's head sticking out of one door and Petro's at the other. Ah, they had, then, some good luck left them. They had had each other in their days of trouble, and now they rested from their building labors and sat happily together in their second home, each with a doorway to enjoy.

伊娃和佩特罗找了一个隐蔽的地方,比其他地方都高,似乎放不下一个歪脖子水罐模样的燕巢。不过,等巢壁砌到需要的高度时,他们就在上面盖了一个小圆顶,只在侧面留了两个口,一个口面向西南,一个口面向东南。这之后再过几天,如果你去他们的那块峭壁下用望远镜望望,也许会看见一个口里是伊娃探出的脑袋,而另一个口里则是佩特罗的脑袋。啊,这么说,他们还是剩有一些好运的。在遇到麻烦的那些日子里,他们相互拥有,不离不弃。现在,在他们的第二个家里,没有了筑巢之累,他们幸福地栖息在一起,享受着各自的出入口。

And later on they had more good luck still. For there came a day when they spent no more time sitting at ease within doors, but flew hither and yon, and then, returning to the nest, clung outside with their tiny feet and stuck their heads in at the open doorway for a brief moment before they were off again. Their nest was too far up for anyone to hear or see what went on within; but there must have been some hungry little mouths yawning all day long, to keep Eve and Petro both so busy hunting the air for insects.

以后,他们的好运还多着呢。因为突然有一天,他们不再悠闲自在地待在巢里了,而是这儿飞飞,那儿飞飞,然后,又回来,用小爪子钩住巢的外面,把脑袋伸进敞开着的门里待一小会儿,然后就又飞走了。它们的巢太高了,里面发生的事情谁也听不见,谁也看不见;不过,一定是几只饥肠辘辘的小燕子,整天张着嗷嗷待哺的小嘴,让伊娃和佩特罗不停地在空中为他们捉虫子。

Soon after this one of the doors was closed, sealed tight with clay. What had happened? Were the little ones inside crowding about too recklessly, so that there was danger of one falling out? Had Eve and Petro come upon an especially good mud–puddle and built a bit more just for the fun of it?

这之后不久,一个门关上了,用黏土给封了个严严实实。怎么回事？是不是里面的小家伙儿们挤来挤去的，太不安分，恐怕会有小燕子掉出去呢？是不是伊娃和佩特罗发现了一个特别好的泥潭，为了好玩，就多砌了一点泥呢？

It was not very many days after this that Eve and Petro and all their comrades ceased coming to the cliff where their curious nests were fastened. Their doorways knew them no more; but over the meadows from dawn till nearly dusk there flew beautiful old swallows bearing upon their foreheads the pale mark of a new moon, and with them were their young.

这之后过不了多少天，伊娃、佩特罗和他们的伙伴就彻底离开峭壁，再也不回他们那令人好奇的泥巢了，巢口再也看不到他们的身影了；不过草场上空，从黎明到黄昏，美丽的成燕们飞来飞去，他们的前额上印着像新月一样浅浅的印记，身边和他们在一起的是他们的孩子们。

At night they sought the marshes, where their little feet might cling to slender stems of bending reeds;

and their numbers were very many.

夜晚，他们就找一处沼泽过夜，小爪子紧紧地抓住被压弯了腰的细芦秆；他们数量非常庞大。

But winter would be coming, and if it still was a long way off, so were the hunting grounds of South America, where they must be flitting away the days when the northern marshes would be frozen over.

可是，冬天即将来临，假如冬天还很远的话，那就说明南美洲的捕食地也很远，因为他们必须在那里度过北方沼泽地结冰的那段日子。

So off they went, Eve and Petro and their young, looking so much like others of the swallow flock that we could not tell who they were, now that they had stopped coming to their nest with one open and one closed doorway.

于是，他们出发了，伊娃、佩特罗和他们的孩子们！他们看上去太像别的燕群了，让我们无法分辨，因为他们现在已经不回那个一个门开着、一个门关着的家了。

They would have far to travel, even if they took the direct over–water route, which many sorts of birds do. But what is distance to Petro, whose strong wings carry him lightly? A mile or a hundred or a thousand even are nothing if the hunting be good. Might just as well be flying south, as back and forth over the same meadow the livelong day, with now and then a rest on the roadside wires, which fit his little feet nearly as well as the reeds of the marsh. Some people think it is for the sake of the hunting that the route of the swallows lies overland, for they fly by day and catch their game all along the way.

这一路路途遥远，即使他们像许多别的鸟那样沿着水上的那条直线飞，也要飞很远。但是对佩特罗来说，距离又算什么？他的翅膀发达有力，完全可以让他飞得轻松自如。如果捕食丰厚，一英里，或一百英里，甚至一千英里都不算什么。最好一路南飞，白天就在一块草地上流连忘返，偶尔落在路边的电线上稍事休息，这些电线和沼泽地里的芦苇一样，用小爪子抓着正合适。有些人认为燕子是为了捕食才选择陆路迁徙的，因为，他们白天飞行，一路都在捕食。

And as they journeyed, Eve and Petro and their flock, south and south and south, maybe the children, here and there, waved their hands to them and called, "Good hunting, little friends of the air, and *good luck* through all the winter till you come back to us a-gain."

他们飞呀飞呀——伊娃、佩特罗和他们的燕群——向南，向南，再向南！也许，偶尔会有孩子们一边挥手一边冲他们喊："捕食顺利，天空中的小朋友！祝你们冬天好运！明年见呀！"

A Famous Landmark
一个著名的地标

6

Uncle Sam

山姆大叔，我们的雄鹰

Uncle Sam stood at the threshold of his home, with an air of dignity. There was enough to fill his breast with honest pride. His home had been a famous landmark for generations before he himself had fallen heir to it. It was the oldest one in the neighborhood. It had stood there seventy-five years before, when a white man had built a cabin within sight of it, for company. That cabin had been neglected and had fallen to bits years ago; but Uncle Sam's ancestors had taken care of their place, and had mended the weak spots each season, and had kept it in such repair that it was still as good as ever. It would last, indeed, with such treatment, as long as the post and the beams that supported it held. The post was the trunk of a tall old tree, and the beams were the branches, so near the top that it would be a very brave or a very foolish man who would try to climb so far, for there were no stairs.

山姆大叔站在自己的家门口，神情威严庄重。他当然可以满怀自豪之情了。在他还没有继承家业之前，他的家世世代代就已经是著名的地标了，堪称当地最古老的地标。七十五年前他就已经矗立在那里了，当时有一个白人在附近盖了一个小木屋，以便相互有个照应。那个小木屋由于年久失修，许多年前就已经七零八落的了。可是，山姆大叔的家却得到了祖先们精心的维护，他们每个季节都会修补那些脆弱的地方，由于修缮及时到位，他们的家依然完好如故。的确，在如此精心的维护下，只要起支撑作用的那根柱子和那些横梁不倒，他的家就能代代相传。那根柱子其实就是一棵树的树干，那些横梁就是树枝，那是棵老树，很高的老树，所以如果有人试图爬到树顶那儿，那么这个人要么就是胆子特别大，要么就是特别愚蠢，因为那里可没有楼梯。

No stairs, and such a distance up! But Uncle Sam could find the path that led to it; for was he not a lord of the air, and could he not sail the roughest wind with those strong wings of his?

没有楼梯,而且还那么高! 可是山姆大叔知道能从哪里上去,要知道他可是空中之王,他的翅膀强劲有力,任狂风肆虐,他都能翱翔自如。

Perhaps it was the sure strength of his wings that gave him a stately poise of pride even as he rested. It could not have been the honor men had bestowed upon him; for, although that was very great, he knew nothing about it.

也许是他那对翅膀毋庸置疑的力量让他即使休息时也能镇定自若、傲然矗立。不可能是因为人类加在他身上的那些光环。尽管这些光环无比辉煌,但是,他对此却一无所知。

Soldiers had gone into battle for freedom and right, bearing the picture of Uncle Sam on their banners. Veterans had walked in Memorial Day parades, while over their gray heads floated the symbol of Uncle Sam and the Stars and Stripes. Yes, the people of a

Above all other creatures of this great land he had been honored.
在这片伟大的国土上，他获得的荣耀至高无上，任何其他生物都望尘莫及。

great and noble land, reaching from a sea on the east to a sea on the west, had honored Uncle Sam by choosing him for the emblem of their country. His picture was stamped on their paper money, and ornamented one side of the coins that came from the mint, with the words, "In God We Trust," on the other side. Above all other creatures of this great land he had been honored; and could he have understood, he might well have been justly proud of this tribute.

士兵们为了自由和正义奔赴战场,他们的旗帜上印着山姆大叔的画像;阵亡将士纪念日的巡游队伍里,山姆大叔的形象和星条旗在老兵们灰白的头上迎风飘扬。是的,这片雄伟壮丽的国土从东面的大海一直延伸到西面的大海,这片国土上的人民选择山姆大叔作为国家的象征,这是赋予他的荣耀。纸币上印着他的形象,铸币厂里出来的硬币上,一面装饰着山姆大叔的头像,另一面写着"我们信仰上帝"。在这片伟大的国土上,他获得的荣耀至高无上,任何其他生物都望尘莫及。假如他能够理解,那么,他一定会为这实至名归的荣耀而感到无比自豪。

But as it was, perhaps his emotions were centred

only on his family; for his home was shared by his mate and two young sons. He bent his white head to look down at his twins. They were such hungry rascals and needed such a deal of care! They had needed care, indeed, ever since the day their little bodies had begun to form in the two bluish white eggs their mother had laid in the nest. They had stayed inside those shells for a month; and they never could have lived and grown there if they had not been brooded and kept warm. Their mother had snuggled her feathers over them and kept them cosy; and, when she had needed a change and a rest, Uncle Sam had cuddled them close under his body; for a month is a long time to keep eggs from getting cold, and it was only fair that he should take his turn.

可尽管如此，他可能还是把所有的情感都倾注到自己的家庭上了，因为他的家里还有妻子和两个儿子。他低下白蒙蒙的脑袋，看着他的双胞胎儿子。简直就是两个贪吃鬼！总是吃不够，照顾他们可真不容易！的确，自从母亲在巢里产下了两枚青幽幽的蛋，他们就在那里蠢蠢欲动了，从他们的小身体开始成形的那一天起，他们就需要照顾了。他俩在蛋壳里一待就是一个月，如果没有孵化和保温，他俩在里面是绝对活不了的，也不可能长大。他们的母亲舒展开全身的羽毛，把他们拢在身下，让他们好舒服呀！需要变变姿势休息一下的时候，山姆大叔就过来把他们紧紧地拥在自己身下。要让两个蛋宝宝在一个月的时间里免受风寒,可不是一件容易的事情。所以,山姆大叔理应轮班孵蛋,这才算公平。

He was no shirk in his family life. He had chosen his mate until death should part them; and whenever there were eggs in the nest, he was as patient about brooding them as she was; for did they not belong to both of them, and did they not contain two fine young eagles in the making?

在家庭生活中,山姆大叔从不逃避责任。他与自己的伴侣终身相依、至死不渝;无论何时,只要巢里有蛋,他都会和妻子一样耐心地孵化他们。蛋是属于他俩的,里面还装着两只可爱的鹰宝宝呢。

And never had they had finer children than the two who that moment were opening hungry mouths and begging for food. In answer to their teasing, Uncle Sam spread his great wings and took stately flight to the lake. For he was a fisherman. When a fish came to the surface, he would try to catch it in his strong claws, so that he might have food to take back to his waiting family. This was easy for him when the fish was wounded or weak and had come to the surface to die; but the quick fishes often escaped, because he was not so skillful at this sort of fishing as the osprey.

此时此刻，这两只鹰宝宝正嗷嗷待哺，这个时候的鹰宝宝是最可爱的。在他们没完没了的纠缠下，山姆大叔张开他那对巨大的翅膀，身姿矫健地向湖那边飞去，因为他是一个渔夫。有鱼浮到水面时，他就想方设法把他们抓到强劲有力的爪子里，这样回家的时候，他就能给期盼的亲人们带去食物啦。如果鱼儿受了伤或者生了病，漂到水面，奄奄一息时，抓住他轻而易举。可是那些行动敏捷的鱼儿常常逃脱，山姆大叔毕竟不善此道，他的捕鱼技巧远不如鱼鹰娴熟。

Yes, the osprey was a wonderful fisherman, who could snatch a fish from the water in his sure claws. But for all that, he was not so wonderful as Uncle Sam, who could catch a fish in the air.

不错，鱼鹰才真是技术精湛的捕鱼高手呢，他的双爪稳健有力，一下子就能抓住水里的鱼。可尽管如此，鱼鹰还是不如山姆大叔棒，因为山姆大叔能在空中捕鱼。

Now, fishing in the air was a thrilling game that Uncle Sam loved. All the wild delight of a chase was in the sport. He used, sometimes, to sit high up on a cliff and watch the osprey swoop down to the water. Then, when the hawk mounted with the prize, Uncle Sam flew far above him and swept downward, commanding him to drop the fish. The smaller bird obeyed, and let the fish fall from his claws. But it never fell far. Uncle Sam closed his mighty wings and dropped with such speed that he caught the fish in mid-air; and the tree-tops swayed with the sudden wind his passing caused. Surely there was never a more exciting way of going fishing than this!

空中捕鱼惊心动魄，是山姆大叔热爱的一项运动，所有疯狂追逐的乐趣尽在其中。有时候，他伫立在高高的悬崖上，看着鱼鹰向水面俯冲下去，当鱼鹰抓着战利品飞离水面，升到空中的时候，山姆大叔就远远地飞到他的上面，然后猛冲下去，逼迫鱼鹰丢掉刚刚捕获的鱼。比他弱小的鱼鹰只好俯首听命，松开爪子，让鱼儿落下。不过他落不了多远，山姆大叔就会收起强劲有力

The Yankee-Doodle Twins
“扬基杜德尔”双胞胎

的翅膀，迅速向下坠落，在半空中将鱼儿收到自己的利爪之下。他飞过之处，掀起一阵疾风，吹得树梢左摇右摆。确确实实，世上再也没有比这更激动人心的捕鱼方式了。

And did the fish belong to the osprey or to Uncle Sam?

那么，这条鱼属于鱼鹰呢？还是属于山姆大叔？

What would you call a man who, by power of greater strength, took away the food another man had earned?

如果一个人，仗着自己的力气比别人大，就强行夺走人家辛辛苦苦挣来的食物，那么你说他是什么人？

Are we, then, to call Uncle Sam a thief and a bully?

我们是否得说山姆大叔是“窃贼”，是“恶棍”呢？

Ah, no; because it is not with an eagle as it is with a man.

哎呀，这可不行！因为它是鹰，和我们人类可不一样。

For the wild things of the world there is only one law, and that is the Law of Nature. They must live as they are made to live, and that is all that concerns them. There is nothing for bird or beast or blossom to learn about "right" or "wrong," as we learn about those things. All they need to do—any of them—is to live naturally.

对于世界上的野生动物来说，只有一条法则，这条法则就是“自然法则”。他们必须生存，这是他们的使命，也是他们唯一关心的事情。无论是鸟儿、兽儿，还是花儿，他们不必像我们一样知道什是“对”，什么是“错”。他们唯一要做的事情就是按照自然法则生存下去。

When we think about it that way, it is very easy to tell whether the fish belonged to the osprey or to

Uncle Sam. Of course, to begin with, the fish belonged to itself as long as it could dive quickly enough or swim fast enough to keep itself free and safe. But the minute the osprey caught it, it belonged to the osprey, just as much as it would belong to you if you caught it with a net or a hook. Yes, the fish belonged to the osprey *more* than it would belong to you; for ospreys hunted food for themselves and for their young in that lake centuries and centuries before a white man even saw it, and before nets and hooks were invented; and besides, in most places, the children of men can live and grow if they never eat a fish, while the children of the osprey would die without such food. So we admire Fisherman Osprey for his strength and swiftness and skill, and are glad for him when he flies off with the prize, which is his very own as long as he can keep it.

这样一想，就不难明白这条鱼到底是属于鱼鹰，还是属于山姆大叔了。当然，首先这条鱼在游得够快、潜得够深，能保持自身自由和安全的情况下，它属于它自己；但是从鱼鹰抓到它的那一刻起，它就属于鱼鹰了，这就和你用渔网网住的鱼或者用鱼钩钓上来的鱼属于你一样，是一回事。其实，虽然这条鱼可以属于你，但它更应该属于鱼鹰，因为早在白人连湖都没见过的无数个世纪之前，鱼鹰们就已经在这个湖里靠捕鱼养活自己和家小了，那个时候还没有发明渔网和渔钩呢。再说了，在大多数地方，人类的孩子即使从不吃鱼也照样能活，照样能长大，可是鱼鹰的孩子们如果没有鱼就会饿死。因此，我们赞叹“捕鱼高手鱼鹰”过人的力量、敏捷的动作和娴熟的技巧，同时，当他带着战利品飞走时，我们也由衷地为他感到高兴，因为只要他能够守得住，这个战利品就属于他。

But when he drops it, it is his no longer, but the eagle´s, who fishes wonderfully in the air—a game depending on the keenness of his sight, his strength, his quickness, and his skill; and the fish that belonged first to itself, and then to the osprey, belonged in the end to the eagle; and all this is according to the Law of Nature.

可是一旦鱼鹰丢掉它，它就不再属于鱼鹰，而是属于山姆大叔这只雄鹰了。山姆大叔在空中的捕鱼过程精彩绝伦，这是一场眼光、力量、动作和技巧的较量。那条本来属于他自己的鱼，后来为鱼鹰所拥有，最后又到了山姆大叔这只雄鹰的利爪之下。所有这一切遵循的都是“自然法则”。

Uncle Sam was not selfish about that fish. He gave it to his twins, and they did enjoy their dinner very, very much, indeed. A fresh brook trout, browned just right, never tasted better to you. For they had been hungry, and the food was good for them.

山姆大叔没有私自占有这条鱼。他把它带给了他的双胞胎儿子，两个小宝宝真的非常、非常喜欢这顿盛宴。这是一条北美溪鳟，身体的褐色不浅不深恰到好处，所以吃起来也不老不嫩正合适，太好吃啦！哥儿俩已经饿了很长时间了，所以这简直就是美味佳肴呀！

Uncle Sam and his mate, whom the children who lived within sight of their nest named Aunt Samantha, had many a hunting and fishing trip to take while the twins were growing; for the bigger the young eagles became, the bigger their appetites were, too. But at last the youngsters were old enough and strong enough and brave enough to take their first flight. Think of them, then, standing there on the outer porch of their great home in the air, and daring to leave it, when it was so very high and they would have so very far to fall if their wings did not work right!

住在鹰巢附近的孩子们把山姆大叔的夫人叫作萨

曼塔大婶，在双胞胎儿子长身体的阶段，她和山姆大叔得不停地出去猎食、捕鱼。两个孩子越长越大，吃得也就越来越多。不过，终于有一天他们长大了，身强体壮了，有勇气开始他们的第一次试飞了。想想当时的情景吧：他们的家结结实实、高高地悬在半空中，他们就站在家门口的最外边，义无反顾地跳了下去！那可是在高高的半空中呀，一旦翅膀无法正常工作，那他们可就跌惨了！

Nonsense, an eagle fall! Had they not been stretching and exercising their muscles for days? And surely the twins would succeed, with Uncle Sam and Aunt Samantha to encourage and urge them forth.

胡说八道！鹰怎么会跌下去呢？你不知道他们一直都在做肌肉拉伸运动吗？他们已经练了好多天了。再说，有山姆大叔和萨曼塔大婶的鼓励和敦促，这对双胞胎哥儿俩的试飞一定会成功的。

The day Uncle Sam cheered his young sons in their baby flight was a great day for all the country round. For not only were the sons of eagles flying, but the sons of men were flying, too. Yes, it was practice day near the lake, and across the water airships rose from the camp and sailed through the air, like mighty birds meant for mighty deeds. For Uncle Sam´s country was at war, and many brave and noble lads thrilled with pride because they were going to help win a battle for Right.

山姆大叔为儿子的初次飞行鼓劲加油的那一天，对整个国家来说意义也非同寻常。这天，不仅鹰的儿子在飞，人类的儿子也在飞。是的，这一天是“湖畔试飞日”。飞艇从营地里腾空而起，越过水面，穿空而过，是呀，大鸟就要有大鸟的身手。山姆大叔的国家正在遭遇战事，许多英勇无畏、品德高尚的小伙子都激动不已，他们将奔赴战场为正义而战，为此他们感到无比的自豪和骄傲！

The bravest and noblest and most fearless of all the camp caught sight of Uncle Sam and smiled. "Emblem of my country! " the young man said. "King of the air in your strong flight! Great deeds are to be done, O Eagle with the snow-white head, and your banner will be foremost in the fight."

营地里最勇敢、最高尚、最无畏的那个青年人看到了山姆大叔，脸上露出了微笑。他由衷地赞叹道："你是祖国的象征！矫健翱翔的天空之王啊！你的丰功伟绩即将实现，噢！头顶白雪的鹰，你的旗帜是战斗的先锋！"

Uncle Sam made no reply. He was too far away to hear, and he could not have understood if he had been near. He saw the distant airships, so big and strong, and led his family away to quieter places, without knowing at all what the big birds were, or what they meant to do. There was so much happening in the country that honored him that Uncle Sam could not understand!

山姆大叔没有回答。他离得太远，听不到年轻人的赞叹。其实，即使他在附近，他也不会明白年轻人说的话。他看见远处的飞艇又大又壮，就带着全家离开了，去找安静一些的地方了，他根本不知道这些大鸟是干什么的，也不知道他们打算去干什么。这个国家正经历着太多的事情，这些事情带给山姆大叔太多的荣誉，可是山姆大叔却无法理解！

He did not even know that, far to the northwest, there was a part of the country called Alaska, where eagles had lived in safety and had brought up their young in peace long after their haunts in most parts of the land had been disturbed. He did not know that the government of Alaska was at that moment paying people fifty cents for every eagle they would kill, and that in two years about five thousand of these noble birds were to die in that manner. He did not know that, if such deeds kept on, before many years there would be no eagles flying proudly through the air: there would be only pictures of eagles on our money and banners. If he could have been told what was happening, and that there was danger that the country would be without a living emblem, and that there might be only stuffed emblems in museums, would he not have thought, "Surely the strong, wise men who go forth to fight for right and liberty will see that the bird of freedom has a home in their land! "

他甚至不知道，离西北很远有一个地方叫阿拉斯加，也是这个国家的一部分。很多年之前，这片国土上鹰赖以生存的栖息地多数遭到破坏，可是，那里的鹰却一直生活得很安全，他们的孩子也一直享受着和平的生长环境；他不知道，此时此刻，那里的人们每杀死一只鹰阿拉斯加政府就会付给他们五十美分，结果，两年之内，大约五千只这种尊贵的鸟儿就这样死于非命；他不知道，如果这种行为继续下去，用不了多少年空中就再也不会有矫健翱翔的鹰了，到那时也许只能在钱币或旗帜上看到他们的画像。假如能把正在发生的一切告诉他，让他知道这个国家的象征即将死去，以后也许只能在博物馆里看到他鼓鼓囊囊的标本，那么他会不会产生这样的想法："那些意志坚强、充满智慧的人们，当他们挺身而出为正义和自由而战的时候，他们肯定会看到在他们的国土上有自由之鸟的一席之地！"

No; Uncle Sam knew nothing about such matters, and so he busied his mind with the things he did know, and was not sad.

不会的，山姆大叔对这些事情一无所知；他把心思都用在了那些他熟悉的事情上，他并不伤心。

He knew where the swamp was, and in the swamp the ducks were thick. They were good-tasting ducks, and there were so many of them that hunters with guns and dogs gathered there from all the country round. And the hunters wounded some birds that the dogs did not get, and these could not fly off at migrating time.

他知道沼泽在哪里，而且沼泽里的鸭子多。那都是些美味的鸭子，而且数量如此众多，以至于猎人拿着枪、带着猎狗从全国各地蜂拥而至。有些鸟逃过了猎狗的追捕，却被猎人的猎枪打伤，迁徙季节到来的时候，受伤的鸟们无法离开了。

Now, Uncle Sam and his family found the wounded ducks easy to catch, and they were nearly as well pleased with them for food as with fish. Of course their feathers had to be picked off first. No eagle would eat a duck with his feathers on, any more than you would. And Uncle Sam knew how to strip off the feathers as well as anyone.

现在，山姆大叔和他的家人发现受伤的鸭子容易捕捉，而且他们喜欢以这些鸭子为食，喜欢的程度几乎不亚于对鱼的喜爱。当然，吃之前，首先要把鸭毛拔掉。鹰不吃带毛的鸭子，正如你不吃带毛的鸭子一样。和所有的人一样，山姆大叔知道如何去掉鸭毛。

So it was interesting in the swamp, and Uncle Sam and Aunt Samantha and the twins were satisfied with hunting there when they were not fishing in the lake.

沼泽地里真是妙趣横生，虽然不在湖中捕鱼了，但是沼泽地里的捕鸭运动也同样让山姆大叔、萨曼塔大婶和他们的双胞胎儿子心满意足。

One day, when Uncle Sam went hunting, he flew near a field where there was a little lamb; and being a strong and powerful eagle, he was able to carry it away. Perhaps he felt very proud as he flew off with so much food at one time. Such strength is something to be pleased with when it is put to the right use, and getting food is as important for an eagle′s life as it is for a man′s.

一天，山姆大叔外出猎食，飞近一块田地时，发现田里有一只小羊。山姆大叔可是一只强壮威猛、力大无穷的鹰，所以抓走小羊对他来说不在话下。一次就能捕获这么一大堆食物，也许让他很自豪吧？用于正道的实力才是值得夸耀的实力。对鹰来说，获取食物至关重要，这关系到他的生死，正如人类需要获取食物才能生存一样。

He lifted his burden high in the air, holding it in his strong talons; and he did not falter once in his steady flight, although the load weighed nearly as much as he did, and he carried it two miles without resting once.

山姆大叔飞到空中，一双强劲有力的爪子紧紧地抓着猎物。虽然小羊的体重几乎和他一样，但是，他飞得很稳当，摇摇欲坠的事情一次也没发生。他带着猎物一口气飞了两英里，中间一次都没休息。

Yes, I think Uncle Sam was proud of that day's hunting and happy with what he had caught; and the tender meat tasted good to him and his family.

这天的捕猎一定让山姆大叔感到自豪了吧？收获的猎物让他很开心吧？鲜嫩的羔羊肉也一定让他们一家子美餐了一顿吧？我想是的。

But the man who had owned the lamb before Uncle Sam caught it was not pleased. He happened to be coming out of the woods just in time to see the capture; and an hour later the boy and the girl who lived within sight of Uncle Sam´s nest met the man and saw that he carried a gun.

可是，小羊以前的主人不高兴了，没被山姆大叔逮住之前，小羊可是属于他的呀。山姆大叔抓小羊时，他正好从树林里出来，看见了那一幕。一个小时后，住在山姆大叔家附近的那个男孩和女孩遇到了他，看见他手里拿着一杆枪。

"I´m after a white–headed sheep thief," he said, "do you know which way he flew, after he reached the cliff?"

“我在追一个白脑袋的偷羊贼，”他说，“你们知道他到悬崖后，又朝哪边飞去了吗？”

The boy´s face turned white in a second, and he held his fists together very still and very tight. The girl looked at her younger brother and then at the man.

霎时，小男孩脸色苍白，他攥起双拳，紧紧地、一动不动。小女孩看看她的弟弟，又看看那个男人。

"Yes, we know," she said, "and we will not tell."

“是的，我们知道，”她说，“可是我们不会说的。”

"Why?" asked the man. "He took the lamb I was going to roast when it was big enough."

“为什么？”那个男人问。“他抓走了我的小羊，等它长大了，我还要烤了吃呢。”

The girl chuckled a little merrily. "And Uncle Sam got ahead of you," she said. "Never mind, I´ll get the money to pay for his dinner. The eagles here usually eat fish from the lake, and sometimes game from the swamp; but once in a very, very long while they take a lamb. When that happens, the Junior Audubon Society at our school pays for their treat. I have the money, because I am treasurer."

小女孩咯咯笑了几声。“可惜山姆大叔抢先了，”她说。“不要紧，我会把钱给你，替他买单的。这里的鹰一般只吃湖里的鱼，有时还会去沼泽那里捕食。偶尔他们也会捉小羊，不过这样的事情很少发生。一旦发生，我们学校的‘少年奥杜邦协会’就会掏钱，算是请他们吃了。我有钱,因为我就是协会的财务主管。”

After the girl turned back to the house for the money, the boy looked hard at the gun. Then he swallowed to get rid of the lump that hurt his throat and said, "If you had shot Uncle Sam or Aunt Samantha or their young, the children for miles and miles NEVER would have liked you. Eagles have nested in that tree for more than seventy years, and nobody except a newcomer would think of shooting one."

小女孩转身回家去取钱了,小男孩则使劲地盯着那杆枪。他咽了一口唾沫,把咔在喉咙里的那团东西吞了下去,然后说道:“假如你杀死了山姆大叔、萨曼塔大婶,或者他们的孩子,方圆几英里、几十英里,甚至几百英里的小孩子就再也不会喜欢你了。鹰在那个树上筑巢七十多年了,除非是陌生人,否则谁也不会想到要射杀他们。”

So they talked together for some time about eagles; and when the girl came back, the man did not charge so much for Uncle Sam´s treat as we sometimes have to pay for our own lamb chops.

于是他们说起了鹰，过了一会儿，那个女孩回来了。结果，山姆大叔的这顿饭钱，那个男人没要多少，就当请客了，有时候，我们自己吃羊排不也要花钱嘛。

And way off among the cliffs Uncle Sam ate in content, not knowing that his life had been in danger, and that he had been saved by a boy and a girl who were growing up "under the shadow of an eagle´s wings," as they said to each other as they watched him sail the air in his journeys to and fro.

远远的悬崖峭壁上，山姆大叔尽情地吃着，根本不知道刚才他的命危在旦夕，也不知道是一个小男孩和一个小女孩救了他的命。正如他俩在一起时曾经说过的那样，他们是在“雄鹰翅膀的庇护下”长大的，说这番话的时候，山姆大叔正在他们眼前的天空中展翅翱翔，一趟趟飞在他的旅途。

That afternoon, when they heard him call, "Cac, cac, cac," they said, "Uncle Sam is laughing." And when his mate answered in her harsh voice, they said,

"Aunt Samantha would be happy if she knew we saved their lives."

那天下午，听到山姆大叔"嘎咯，嘎咯，嘎咯"的大叫时，他们说："山姆大叔在哈哈大笑呢。"听到萨曼塔沙哑的回答时，他们说："假如萨曼塔大婶知道我们救了他们的命，一定会高兴的。"

Busy with the life Nature taught them to live, the twins grew up as Uncle Sam had grown before them.

大自然妈妈教会了两只小鹰如何去生活，和山姆大叔一样，他们整天忙忙碌碌的，不知不觉间就长大了。

As they were hunters, there was nothing more interesting to them than seeking their food in wild, free places. They had no guns and dogs, but they caught game in the swamp. They had no cooks to prepare their ducks, so they picked off the feathers themselves. They had no fish-line and tackle, but they caught fish in the lake. And in time they caught fish in the air, too; which was even more thrilling, and a game they came to enjoy when they overtook the ospreys. Many times, too, they sought the fish that had been washed up on the lake shore, and so helped keep things sweet and clean. In this way they were scavengers; and it is always well to remember that a scavenger, whether he be a bird or beast or beetle, does great service in the world for all who need pure air to breathe.

他们是猎手,对他们来说,干什么都不如在荒郊野外自由自在地猎食有趣。没有猎枪,也没有猎狗,他们照样能在沼泽里捕获猎物;没有厨师为他们收拾捕获的鸭子,他们就自己动手拔去鸭毛;没有渔线和渔具,他们照样能捉到湖里的鱼。时机赶得好,他们还能在空中捕鱼呢!空中捕鱼更惊心动魄,和鱼鹰较量的成功让他们喜欢上了这种运动。很多时候,他们也会在湖边寻觅被湖水冲上岸的鱼,所以,湖边的清新和洁净也有他们的功劳。这么说来,他们也算是食腐动物了。我们应该时刻牢记:无论是鸟,也无论是野兽或者是甲壳虫,只要他是食腐生物,在这个世上,他就为需要纯净空气的万物做出了巨大的贡献。

The first year they became bigger than their father, and bigger than they themselves would be when they were old. At first, too, their eyes were brown, and not yellow like their father´s and mother´s. And for two years their heads and tails were dark, so that they looked much more like "golden eagles" than they did like the old ones of their own kind.

仅一年的工夫,他们的个头就超过了父亲,而且一直到老都不再长了，那时恐怕只有缩的份了。父亲、母亲的眼睛是黄色的，可是他们的眼睛刚开始却是棕色的。脑袋和尾巴的颜色头两年都很深,结果,看上去他们更像“金雕”,而不像自己的长辈。

The soldiers at the training–camp caught sight of them now and then, and named them the "Yankee–Doodle Twins", when the twins were three years old, their molting season brought a remarkable change to them. The dark feathers of their heads and necks and tails dropped out, and in their places white feathers grew, so that by this time they looked like their own father and mother, who are what is called "bald eagles", though their heads are not bald at all, but well covered with feathers.

训练营里的士兵们时不时地看见他们,于是就给他们起了个名字:“扬基督德双胞胎”(“扬基督德”是一首歌的歌名,美国独立战争期间,这首歌在美国士兵中非常流行，现在号称是美国的第二国歌)。三岁那年的换毛季节，这对双胞胎身上发生了巨大的变化。他们脑袋、脖子和尾巴上的深色羽毛全掉了，取而代之的是白色的羽毛，直到此时，他们看上去才像自己的父母亲了，他们变成了人们说的那种“秃鹰”，可是他们的脑袋不仅一点儿都不秃，还长着厚厚的羽毛呢。

These two birds that were hatched in the home that was more than seventy years old lived to see the end of the war the young soldiers were training for when they took their first flights together near the shore of the same lake. And perhaps they will live to a time when the people of their country learn to deal more and more justly with each other and with the great bird of freedom chosen by their forefathers to be the emblem of their proud land.

在有七十多年历史的老屋里出生的这对鹰亲眼见证了那场战争的结束——那场他俩初次在湖边试飞的时候，那些年轻的士兵为之训练的战争；也许当这片国土上的人们学会了如何与同类和平相处，学会了如何与祖先为他们为之骄傲的国土选择的自由之鸟和平相处的时候，他们依然能活着见证这一切。

Why, indeed, if the boys and girls of the neighborhood keep up a guard for the protection of Uncle Sam and Aunt Samantha, should they not nest again, and yet again, in that tree-top home that has been so well taken care of for more than threescore years and ten; and bring up Yankee-Doodle Twins for their country in days of peace as they did in days of war?

的确,为什么不呢?假如周围的那些小男孩和小女孩能时时刻刻提高警惕,保护山姆大叔和萨曼塔大婶,他们肯定会在那个精心维护了七十多年的树顶小屋里世世代代繁衍下去,为祖国精心哺育“扬基督德双胞胎”,战争年代如此,和平时期也如此。

7

Corbie
乌鸦考比

Corbie's great-great-grandfather ruled a large flock from his look-out throne on a tall pine stump, where he could see far and wide, and judge for his people where they should feed and when they should fly.

考比的爷爷的爷爷的爷爷是一国之主，统治着一大群乌鸦。他的御座就在一个高高的松树桩上，那也是他的瞭望台。他坐在那里高瞻远瞩、君临天下，为他的子民们寻找着觅食的地点和时机。

His great-grandfather was famous for his collections of old china and other rare treasures, having lived in the woods near the town dump, where he picked up many a bright trinket, chief among which was an old gold-plated watch-chain, which he kept hidden in a doll's red tea-cup when he was not using it.

考比的爷爷的爷爷赫赫有名。他曾经住在镇上垃圾场附近的树林里，所以，收藏了大量的旧瓷器和别的稀世珍宝，还捡了许多亮闪闪的小玩意儿。其中最珍贵的是一条旧表链，上面还镀着一层金。不戴的时候，他总是把它藏在一个洋娃娃的红茶杯里。

His grandfather was a handsome fellow, so glistening that he looked rather purple when he walked in the sunshine; and he had a voice so sweet and mellow that any minstrel might have been proud of it, though he seldom sang, and it is possible that no one but Corbie's grandmother heard it at its best. He was, moreover, a merry soul, fond of a joke, and always ready to dance a jig, with a chuckle, when anything very funny happened in crowdom.

他的爷爷是一个帅小伙儿。他浑身闪闪发亮，在阳光下漫步时，散发出紫色的光辉。他嗓音柔和悦耳，富有磁性，令每一位吟游歌手都为之倾倒。不过他很少歌唱，他那美妙的歌声恐怕只有考比的奶奶才有幸听到过。另外，他天性快乐，非常喜欢开玩笑；每当乌鸦王国里发生什么好笑的事情，他都要蹦上蹦下地跳上一曲吉格舞，一边舞还一边咯咯地笑个不停。

As for the wisdom and beauty of his grandmothers all the way back, there is so much to be said that, if I once began to tell about them, there would be no space left for the story of Corbie himself.

至于考比奶奶们的智慧和美貌，要是一代一代地往上倒腾，那要说的可就太多了，一旦打开话匣子，恐怕就没法再讲考比自己的故事了。

Of course, coming from a family like that, Corbie was sure to be remarkable; for there is no doubt at all that we inherit many traits of our ancestors.

当然，出生在这样一个家族，考比也一定非同凡响，因为毫无疑问，我们都会继承祖先的许多特征。

In this Mother Crow had laid her eggs.
乌鸦妈妈把自己的蛋产在了这里。

Corbie knew very little about his own father and mother, for he was adopted into a human family when he was ten days old, and a baby at that age does not remember much.

考比对自己的爸爸妈妈所知甚少，因为他出生刚刚十天，就被人类收养了，那么小的宝宝是记不了太多事情的。

Although he was too young to realize it, those first ten days after he had come out of his shell, and those before that, while he was growing inside his shell, were in some ways the most important of his life, for it was then that he needed the most tender and skillful care. Well, he had it; for the gentleness and skill of Father and Mother Crow left nothing to be desired. They had built the best possible nest for their needs by placing strong sticks criss-cross high up in an old pine tree. For a lining they had stripped soft stringy bark from a wild grapevine, and had finished off with a bit of still softer dried grass.

尽管当时他还太小，还意识不到这些，但是，出壳后的前十天，以及之前在蛋壳里生长发育的那些日子，在某种程度上是考比一生中最重要的日子了，因为那段日子正是考比最需要细心照料和精心呵护的时期。还好，这一切考比都得到了。乌鸦爸爸和乌鸦妈妈对他的呵护是那样的细致入微、温存体贴，简直让他别无所求。他们找来结实的小棍子，交叉着放在一棵高高的老松树上，为他们建造了一个最好的窝，一个最能满足他们需要的窝。他们又从野葡萄藤上剥来一缕一缕软软的藤皮，垫在窝里，最后又在上面铺了一些比藤条皮还软的干草。

In this Mother Crow had laid her five bluish-green eggs marked with brown; and she and Father Crow had shared, turn and turn about, the long task of keeping their babies inside those beautiful shells warm enough so that they could grow.

乌鸦妈妈就在这里产下了五枚青绿色的蛋，蛋上还带着褐色的斑点，漂亮的蛋壳里面是他们的五个宝宝。为了他们的生长发育，她和乌鸦爸爸得轮流倒换着共同为他们保暖，这可是一项漫长而艰巨的任务呀。

And grow they did, into five as homely little objects as ever broke their way out of good-looking eggshells. There was not down on their bodies to make them fluffy and pretty, like Peter Piper´s children. They were just sprawling little bits of crow–life, so helpless that it would have been quite pitiful if they had not had a good patient mother and a father who seemed never to get tired of hunting for food.

是呀，他们在生长、在发育，最后变成了五只并不漂亮的小东西，和所有啄破漂亮的蛋壳来到这个世界上的小东西一样，都不算漂亮。他们身上没有软软的细绒毛，所以看上去不像矶鹞彼得的孩子毛茸茸的很漂亮。他们只不过是几只刚会爬来爬去的小乌鸦，一副无助的样子，如果没有慈祥、耐心的妈妈和爸爸不知疲倦地为他们觅食，简直就是几只小可怜儿。

Now, it takes a very great deal of food for five young crows, because each one on some days will eat more than half his own weight and beg for more. Dear, dear! how they did beg! Every time either Father or Mother Crow came back to the nest, those five beaks would open so wide that the babies seemed to be yawning way down to the end of their red throats. Oh, the food that got stuffed into them! Good and nourishing, every bit of it; for a proper diet is as important to a bird baby as to a human one. Juicy caterpillars—a lot of them: enough to eat up a whole berry-patch if the crows hadn´t found them; nutty-flavored grasshoppers—a lot of them, too; so many, in fact, that it looked very much as if crows were the reason the grasshoppers were so nearly wiped out that year that they didn´t have a chance to trouble the farmers´ crops; and now and then a dainty egg was served them in the most tempting crow-fashion, that is, right from the beak of the parent.

唉,五只小乌鸦需要的食物还真不少呢。有一段时间,一只小乌鸦就会吃掉比自己体重的一半还重的食物,这还不够,还要张嘴要呢。天哪,天哪!瞧他们那讨吃的样子!每次无论是乌鸦爸爸还是乌鸦妈妈一回到家,那五只嘴巴就张得大大的,就好像是五个宝宝在使劲儿地打哈欠,一直打到露出了他们那红红的嗓子眼儿。噢,瞧那些填进他们肚子里的食物!每一丁点儿都营养丰富、有益健康!合理的膳食对人类的宝宝很重要,对鸟宝宝来说也同样重要。美味多汁的毛毛虫,很多很多。如果不是被乌鸦捉来吃掉,这么多的毛毛虫足以啃掉一大片浆果林;还有味道像坚果的蚱蜢,也很多很多,实际上,太多了,结果那年好像就是因为乌鸦,蚱蜢几乎都被消灭干净了,让他们没机会去祸害农民的庄稼；时不时地，还会有一颗美味的蛋被送到小乌鸦的嘴里，而且是以最有诱惑力的方式送过来的，对了，是从爸爸和妈妈的嘴巴里直接送到他们的嘴里。

For, as you no doubt have heard, a crow thinks no more of helping himself to an egg of a wild bird than we do of visiting the nests of tame birds, such as hens and geese and turkeys, and taking the eggs they lay. Of course, it would not occur to a crow that he didn´t have a perfect right to take such food for himself and his young as he could find in his day´s hunting. Indeed, it is not unlikely that, if a crow did any real thinking about the matter, he might decide that robins and meadowlarks were his chickens anyway. So what the other birds would better do about it is to hide their nests as well as ever they can, and be quiet when they come and go.

你肯定听说过吧，乌鸦会毫不客气地拿走野鸟的蛋，而且一点儿也不脸红，就像我们人类到鸡舍、鹅舍和火鸡舍等禽舍里拿走鸡蛋、鹅蛋和火鸡蛋时一样理所当然、毫无顾忌。当然啦，白天觅食的时候，乌鸦会发现野鸟蛋，可是，他却不会想到自己完全没有权利把这些野鸟蛋据为己有，并且带回家给自己的孩子们吃。的确，也不是没有这种可能，一只乌鸦确实好好想过这件事情，他最后还是觉得知更鸟和野云雀就是给自己下蛋的鸡。所以，其他的鸟儿所能做的，就是尽可能地把自己的巢藏起来，而且进出的时候要静悄悄的，不出一点儿声息。

That is the way Father and Mother Crow did, themselves, when they built their home where the pine boughs hid it from climbers below and from fliers above. And, though you might hardly believe it of a crow, they were still as mice whenever they came near it, alighting first on trees close by, and slipping up carefully between the branches, to be sure no enemy was following their movements. Then they would greet their babies with a comforting low "Caw," which seemed to mean, "Never fear, little ones, we´ve brought you a very good treat." Yes, they were shy, those old crows, when they were near their home, and very quiet they kept their affairs until their young got into the habit of yelling, "Kah, kah, kah," at the top of their voices whenever they were hungry, and of mumbling loudly, "Gubble-gubble-gubble," whenever they were eating.

乌鸦爸爸和乌鸦妈妈自己就是这样，他们把巢藏在了松树枝之间，无论是谁，不管是从树下爬上来的，还是从上空飞来的，都看不见。而且，也许你不相信，快到家门口时，乌鸦也会像老鼠那样蹑手蹑脚，没有一点儿声息，他们先落到附近的树上，小心翼翼地在树枝间穿梭，溜到上面，看看有没有尾随而来的敌人。确信没有之后，才会“呱”的一声向孩子们打招呼，声音低低的、非常亲切，好像在说：“别怕，小家伙们，爸爸妈妈给你们带好吃的来了。”的确，快到家的时候，老乌鸦们都在担惊受怕，他们非常安静，一举一动都轻手轻脚的。这种情况直到后来才有所改变，因为，后来小乌鸦们养成了一个坏习惯，他们一饿就扯着嗓子大喊“嘎，嘎，嘎”，一吃饭就大声嘟囔：“嘎饱，嘎饱，嘎饱。”

After that time comes, there is very little quiet within the home of a crow; and all the world about may guess, without being a bit clever, where the nest is. A good thing it is for the noisy youngsters that by that time they are so large that it does not matter quite so much.

那时，乌鸦的家里就别想再有一刻的安宁了；周围无论是谁，哪怕是傻瓜，都知道他们的巢在哪儿。好在到了那个时候，这些整天大喊大叫的小乌鸦已经长大了，就算知道也无所谓。

But it was before the "kah-and-gubble" habit had much more than begun that Corbie was adopted; and the nestlings were really as still as could be when the father of the Brown-eyed Boy and the Blue-eyed Girl climbed way, way, way up that big tree and looked into the round little room up there. There was no furniture—none at all. Just one bare nursery, in which five babies were staying day and night. Yet it was a tidy room, fresh and sweet enough for anybody to live in; for a crow, young or old, is a clean sort of person.

可是，考比刚刚养成“嘎——饱”叫的习惯还没几天，就被收养了。那天，几只小乌鸦拼命地屏住呼吸，一动也不敢动，那个“棕眼睛男孩”和那个“蓝眼睛女孩”的爸爸顺着那棵大树爬上来了，他爬得很高、很高，探头朝圆圆的小屋里张望。里面没有家具，一件也没有，只有一个光秃秃的育儿室，五个小宝宝白天黑夜都待在里面，不过，小屋很整洁，也没有味，住起来还算舒适，因为无论是小乌鸦还是老乌鸦，都爱干净。

The father of the Brown-eyed Boy and the Blue-eyed Girl looked over the five homely, floundering little birds, and, choosing Corbie, put him into his hat and climbed down with him. He was a nimble sort of father, or he never could have done it, so tall a tree it was, with no branches near the ground.

"棕眼睛男孩"和"蓝眼睛女孩"的爸爸仔细看了看那五只不知所措，也不算漂亮的鸟儿，最后选定了考比。他把考比放到自己的帽子里，带着它爬到了树下。他是个身手敏捷的爸爸,要不然他永远也无法爬到树上拿走小考比,因为那棵树太高了,靠近地面的地方连一根树枝都没有。

Corbie, even at ten days old, was not like the spry children of Peter Piper, who could run about at one day old, all ready for picnics and teetering along the shore. No, indeed! He was almost as helpless and quite as floppy as a human baby, and he needed as good care, too. He needed warmth enough and food enough and a clean nest to live in; and he needed to be kept safe from such prowling animals as will eat young birds, and from other enemies. All these things his father and mother had looked out for.

虽然考比已经出生十天了，可是一点儿都不活泼，不像矶鹞彼得那些孩子，刚出壳一天就能到处乱跑了，还时刻准备着去野餐和摇来晃去地在岸边漫步。可不是吗，一点儿都不像！他和人类的小婴儿一样，几乎没有自理能力，而且也是一副软塌塌的样子，也需要悉心呵护。他需要温暖，需要食物，需要干净的巢，他还需要保护，保护他不受那些四处潜行偷吃小鸟的动物和天敌的伤害，这可都是考比的爸爸妈妈一直以来严加防范的事情呀。

Now the little Corbie was kidnaped—taken away from his home and the loving and patient care of his parents.

现在小考比被绑架了——被人从家里带走，再也没有爸爸妈妈那无微不至的悉心呵护了。

But you need not be sorry for Corbie—not very. For the Brown-eyed Boy and the Blue-eyed Girl adopted the little chap, and gave him food enough and warmth enough and a chance to keep his new nest clean; and they did it all with love and patience, too.

不过你不必为考比难过，不必太难过，因为那个“棕眼睛男孩”和“蓝眼睛女孩”收养了这个小家伙，给他喂食，为他保暖，还给他机会把自己的新巢弄干净，做这一切时，他们也是那么充满爱怜、那么细致入微。

Corbie kept them busy, for they were quick to learn that, when he opened his beak and said, "Kah," it was meal-time, even if he had had luncheon only ten minutes before. His throat was very red and very hollow, and seemed ready to swallow no end of fresh raw egg and bits of raw beef and earthworms and bread soaked in milk. Not that he had to have much at a time, but he needed so very many meals a day. It was fun to feed the little fellow, because he grew so fast and because he was so comical when he called, "Kah".

考比可把两个孩子忙活坏了，他们很快就知道，只要他张开嘴巴叫“嘎”，就是要吃饭啦，哪怕十分钟前他刚刚才加过一顿餐。它的喉咙红红的，很深，好像随时都能吞下去无数个鲜生蛋、无数块生牛肉、无数条蚯蚓，还有无数片浸过牛奶的面包似的。倒不是他一次就得吃很多，而是他一天要吃好多次，太多次了。给这个小家伙喂食很好玩，因为他长得飞快，而且他张嘴叫

“嘎”的时候，模样滑稽极了。

It was not long before his body looked as if he had a crop of paint–brushes growing all over it; for a feather, when it first comes, is protected by a little case, and the end of the feather, which sticks out of the tip of the case, does look very much like the soft hairs at the end of a paint–brush, the kind that has a hollow quill stem, you know. After they were once started, dear me, how those feathers grew! It seemed no time at all before they covered up the ear–holes in the side of his head, and no time at all before a little bristle fringe grew down over the nose–holes in his long horny beak.

过不了过久，他全身上下就好像长了一层密密麻麻的油彩笔，因为羽毛刚长出来的时候，上面套着一个小护套，羽毛尖从护套尖那里戳出来，看上去的确很像油彩笔尖上那软软的毛，你知道的，就是那种空心杆的油彩笔。这些羽毛一旦开长，噢，我的天哪，长得太快了！好像眨眼之间就把他脑袋两侧的耳朵眼儿给盖起来了；好像眨眼之间，他那长长的角质喙上的鼻孔就被一小圈硬刷刷的短毛给遮住了。

He was nearly twenty days old before he could stand up on his toes like a grown-up crow. Before that, when he stood up in his nest and "kahed" for food, he stood on his whole foot way back to the heel, which looks like a knee, only it bends the wrong way. When he was about three weeks old, however, he began standing way up on his toes, and stretching his leg till his heels came up straight. Then he would flap his wings and exercise them, too.

出生快二十天后，他才能像一只大乌鸦那样用脚趾站立起来。这之前，在窝里“嘎，嘎”地叫着讨食吃的时候，他双脚从前面一直到脚后跟全部着地才能站起来，他的脚看起来好像是人类的膝盖，只不过他的“膝盖”弯错方向了。不过，出生大约三个星期后，他就能脚趾着地站得直直的了，这时，他的双腿拉得长长的，连脚后跟都绷直了。接下来，他会拍拍翅膀，开始练习飞翔的动作。

Of course, you can guess what that meant. It meant—yes, it meant that Corbie was getting ready to leave his nest; and before the Brown-eyed Boy and the Blue-eyed Girl really knew what was happening, Corbie went for his first ramble. He stepped out of his nest–box, which had been placed on top of a flat, low shed, and strolled up the steep roof of the woodshed, which was within reach. There he stood on the ridge–pole, the little tike, and yelled, "Caw," in almost a grown–up way, as if he felt proud and happy. Perhaps he did for a while. It really was a trip to be proud of for one´s very first walk in the world.

当然,你知道这意味着什么了。这意味着——是的,意味着考比要离巢了。那个“棕眼睛男孩”和“蓝眼睛女孩”还没弄明白是怎么回事呢,考比就已经出发了,开始了他的第一次漫游。他迈出巢箱,那个巢箱一直放在一个平坦、低矮的棚顶上,然后,不慌不忙跨到咫尺之遥的那个柴火棚的棚顶上,信步登上了陡峭的棚顶。他站在棚脊的横梁上,这个小淘气,像大乌鸦一样煞有介事地大叫了一声:“嘎”,好像有说不出的自豪和幸福似的。也许那一刻他真的很自豪,也很幸福。毕竟,这是他第一次来到外面的世界,当然值得骄傲,值得高兴了。

But the exercise made him hungry, and he soon yelled, "Kah! " in a tone that meant, "Bring me my luncheon this minute or I´ll beg till you do."

不过,这一趟下来,考比饿了。不一会儿,他就“嘎!

嘎！”地叫了起来，听口气好像是在说：“快把饭拿来，要不我就一直叫！不拿来我就不停！”

The Brown-eyed Boy took a dish of bread and milk to the edge of the low roof, where the nest-box had been placed, and the Blue-eyed Girl called, "Come and get it, Corbie."

“棕眼睛男孩”端了一份面包和牛奶放到低低的屋檐上，巢箱曾经就放在那里，“蓝眼睛女孩”喊着：“过来吃吧，考比！”

Not Corbie! He had always had his meals brought to him. He liked service, that crow. And besides, maybe he *couldn't* walk down the roof it had been so easy to run up. Anyway, his voice began to sound as if he were scared as well as hungry, and later as if he were more scared than hungry.

让考比过来？那可不行！以前都是别人把饭送到他面前的。他喜欢被人伺候，唉，这乌鸦！再说了，上这个斜顶很容易，可是下去就不一定了。总之，考比的声音听上去好像不光是饿了，还有点害怕的意思，再到后来，听上去与其说是他饿了，还不如说是他害怕了。

Now it stood to reason that Corbie´s meals could not be served him every fifteen minutes on the ridge-pole of a steep roof. So the long ladder had to be brought out, and the crow carried to the ground and advised to keep within easy reach until he could use his wings.

这下什么都不用说了，棚顶那么陡，横梁又那么高，根本没有办法每隔十五分钟就伺候考比吃一次饭，所以，不得不请出那架长长的梯子，把考比拿回到地面上，然后警告他在没有学会用翅膀之前，最好还是老老实实地待在大家够得着“伺候”他的地方。

It was only a few days until Corbie could fly down from anything he could climb up; and from that hour he never lacked for amusement. Of course, the greedy little month-old baby found most of his fun for a while in being fed. "Kah! Kah! Kah! " he called from sun-up to sun-down, keeping the Brown-eyed Boy and the Blue-eyed Girl busy digging earthworms and cutworms and white grubs, and soaking bread in milk for him. "Gubble-gubble-gubble," he said as he swallowed it—it was all so very good.

不过,过不了几天,考比就能随心所欲地飞了,爬到什么东西上都不怕了。从那一刻起,他就永远告别了寂寞和无聊,整天乐趣无穷。当然了,有一段时间,这只贪吃的、只有一个月大的小家伙最大的乐趣还是别人给他喂食吃。"嘎!嘎!嘎!"他从日出叫到日落,那个"棕眼睛男孩"和那个"蓝眼睛女孩"忙不迭地为他挖蚯蚓、挖切根虫、挖蛴螬,还要把面包放在牛奶里泡给他吃。"嘎饱——嘎饱——嘎饱,"考比一边吞,一边大声叫着,哇!简直太惬意了!

The joke of it was that Corbie, even then, had a secret—his first one. He had many later on. But the very first one seems the most wonderful, somehow. Yes, he could feed himself long before he let his foster brother and sister know it; and I think, had he been a wild crow instead of a tame one, he would have fooled his own father and mother the same way—the little rascal.

可笑的是,早在那个时候考比就有了自己的秘密,他的第一个秘密。以后,他还会有很多秘密,不过,这第

"Kah! Kah! Kah! " he called from sun–up to sun–down.
“嘎！嘎！嘎！”它从日出叫到日落。

一个秘密似乎最奇妙。是呀，他早就能自己吃饭了，可是收养他的那对兄妹很久以后才知道这个秘密。我想，如果他不是家养乌鸦，而是一只野乌鸦的话，他一定也会把自己的爸爸妈妈骗得团团转的，真是个小淘气鬼！

No one would think, to see him with beak up and open, and with fluttering wings held out from his sides, that the little chap begging "Kah! kah! kah! " was old enough to do more than "gubble" the food that was poked into his big throat. But for all that, when the Brown-eyed Boy forgot the dish of earthworms and ran off to play, Corbie would listen until he could hear no one near, and then cock his bright eye down over the wriggling worms. Then, very slyly, he would pick one up with a jerk and catch it back into his mouth. One by one he would eat the worms, until he wanted no more; and then he would hide the rest by poking them into cracks or covering them with chips, crooning the while over his secret joke. "There-there-tuck-it-there," was what his croon sounded like; but if the Brown-eyed Boy or the Blue-eyed Girl came near, he would flutter out his wings at his sides and lift his open beak, his teasing "Kah" seeming to say, "Honest, I haven´t had a bite to eat since you fed me last."

看着考比嘴巴冲上大张着，扑棱着身体两侧的翅膀，一个劲儿地“嘎，嘎，嘎”地讨食吃，谁也不会想到这个小家伙已经长大了，除了能把送到大喉咙里的食物大口大口地吞下去之外，他还有别的本事呢。可虽然如此，当那个“棕眼男孩”丢下那盘蚯蚓，跑到一边去玩的时候，考比还是会侧耳细听，直到确信附近没有人的时候，才会垂下亮晶晶的眼睛去看那些蠕动的虫子。然后，他才会鬼鬼祟祟地伸出嘴巴，猛地一下啄住一条，吃进嘴里。他吃了一条又一条，直到再也吃不下为止。然后，考比会把剩下的蚯蚓塞到缝隙里藏起来，或者用小木片把他们盖住，整个过程中，他都在为自己的秘密把戏而得意扬扬地哼着小曲儿。他的小曲儿听起来就像是：“那儿，那儿，把他藏在那儿。”不过，一旦那个“棕眼男孩”和那个“蓝眼女孩”来到跟前，他就立即扑棱起两侧的翅膀、抬起张得大大的嘴巴，“嘎，嘎”地纠缠不休，似乎在说：“天地良心！自你上次喂完我之后，我可是连一口都没吃呀！”

When his body was grown so big with his stuffing that he was almost a full-sized crow, he stopped his constant begging for food. The days of his greed were only the days of his growth needs, and the world was too full of adventures to spend all his time just eating.

塞进那么一大堆东西，把肚子撑得鼓鼓的，他看起来几乎就是个大乌鸦了，此时的考比不再喋喋不休地讨食吃了。那段日子他之所以那么贪吃只是因为他需要长身体。这个世界充满了刺激，他要去冒险，可不能把时间都用在吃饭上。

It was now time for him to take pleasure in his sense of sight, and for a few weeks he went nearly crazy with joy over yellow playthings. He strewed the vegetable garden with torn and tattered squash-blossoms—gorgeous bits of color that it was such fun to find hidden under the big green leaves! He strutted to the flower-garden, and pulled off all the yellow pansies, piling them in a heap. He jumped for the golden buttercups, nipping them from their stems. He danced for joy among the torn dandelion blooms he threw about the lawn. For Corbie was like a human baby in many ways. He must handle what he loved, and spoil it with his playing.

现在他要充分享受视觉带来的快乐了。一连几个星期，他都在玩黄颜色的东西，几乎都玩疯了。他把黄灿灿的南瓜花啄碎，撒得菜园里到处都是，那颜色太漂亮了，这儿一点，那儿一块地藏在大大的绿叶子下面，找到它们可真是太好玩了！他大摇大摆地来到花园里，把所有三色紫罗兰上的黄色花瓣都扯下来，堆成一堆；他又蹦又跳地使劲够着毛茛花，把它们从茎秆上咬下来；他把蒲公英花瓣撒在草坪上，然后在这些碎花瓣中兴奋地手舞足蹈。哎，在很多方面考比就和人类不懂事的小孩子差不多，只要他喜欢的东西，都要拿过来玩一玩，直到把它玩坏。

Perhaps Corbie inherited his dancing from his grandfather. It may have come down to him with that old crow´s merry spirit. Whether it was all his own or in part his grandfather´s, it was a wonderful dance, so full of joy that the Brown-eyed Boy and the Blue-eyed Girl would leave their play to watch him, and would call the Grown-Ups of the household, that they, too, might see Corbie´s "Happy Dance".

或许考比跳的舞是从他爷爷那里继承来的，连同老爷子的快乐天性也一并继承了来。总之，不管全部是他自创的，还是他从爷爷那里继承了一部分，他的舞跳得还真不错，欢快极了，连“棕眼睛男孩”和“蓝眼睛女孩”都不玩了，过来看他跳舞，他们还把家里的大人叫过来，让他们也看看考比的“快乐舞”。

If he was pleased with his cleverness in hiding some pretty beetle in a crack and covering it with a chip, he danced. If he spied the shiny nails in the tool-shed, he danced. If he found a gay ribbon to drag about the yard, he danced. But most and best he danced on a hot day when he was given a bright basin of water. Singing a lively chattering tune, he came to his bath. He cocked one bright eye and then the other over the ripples his beak made in the water. Plunging in, he splashed long, cooling flutters. Then he danced back and forth from the doorstep to his glistening pan, chattering his funny tune the while.

当他把漂亮的甲虫藏到缝隙里、用一块碎木片盖住，然后对自己的小聪明沾沾自喜的时候，他就会跳起舞来；当他发现那些在工具篷里闪闪发亮的蜗牛时，他就会跳起舞来；当他看到艳丽的知更鸟在院子里拖着尾巴走来走去的时候，他也会跳起舞来。可是，如果大热天给他一盆清凉的水，那他就跳得最多，跳得也最好。他会叽叽喳喳地唱着轻快的小曲儿，一头扎到自己的澡盆里。然后，他先瞪起一只亮晶晶的眼睛，又瞪起另一只亮晶晶的眼睛，看着水面被他的嘴巴弄出的涟漪。接着，他就跳进去，在里面扑腾好一会儿，水花四溅，啊，真凉爽呀！扑腾够了，他就跳起舞来，一会儿跳到门前的台阶上，一会儿又跳到那盆盈盈的澡盆跟前，一边跳，一边还叽叽喳喳地唱着他那好玩的小曲儿。

Have you heard of a Highland Fling or a Sailor's Hornpipe? Well, Corbie's Happy Dance was as gay as both together, when he jigged in the dooryard to the tune of his own merry chatter. The Brown-eyed Boy and the Blue-eyed Girl laughed to see him, and the Grown-Ups laughed. And even as they laughed, their hearts danced with the little black crow—he made them feel so very glad about the bath. For he had been too warm and was now comfortable. The summer sun on his feathered body had tired him, and the cooling water brought relief. "Thanks be for the bath. O bird, be joyful for the bath! " he chattered in his own language, as he spread his wings and gave again and yet again his Happy Dance.

你听说过苏格兰高地舞，或者水手跳的号笛舞吗？嗯，这两种舞加起来也不如考比在门前院子里随着自己那叽叽喳喳的欢快曲子跳的“快乐舞”欢快呢。看着他那副样子，“棕眼睛男孩”和“蓝眼睛女孩”笑了，那些大人也笑了。大家笑的时候，甚至连心都随着这只黑黑的小乌鸦欢快地跳起来了，他们好高兴呀，看来这个澡洗得真不错！刚才太热了，现在总算舒服了。夏天的烈日照在他满身羽毛的身上，让他没精打采的，是凉爽的水让他立刻神清气爽起来。“谢谢，谢谢给我洗澡。噢，鸟儿啊，为这个凉爽的澡欢呼吧！雀跃吧！”他用自己的语言喋喋不休地道着谢，一边道谢，一边还张开翅膀，一遍又一遍地表演起了他的“快乐舞”。

But a basin, however bright, is not enough to keep a crow in the dooryard; for a crow is a bird of adventure.

可是，一个澡盆，无论里面的水有多清亮，也无法拴住乌鸦的心，让他一直待在院子里，因为乌鸦天性喜欢冒险。

So it was that on a certain day Corbie flew over the cornfield and over the tree-tops to the river; and so quiet were his wings, that the Brown–eyed Boy and the Blue–eyed Girl did not hear his coming, and they both jumped when he perched upon a tiny rock near by and screamed, "Caw," quite suddenly, as one child says, "Boo," to another, to surprise him. Then the bird sang his chatter tune, and found a shallow place near the bank, where he splashed and bathed. After that, the Blue–eyed Girl showed him a little water–snail. He turned it over in his beak and dropped it. It meant no more to him than a pebble. "I think you´ll like to eat it, Corbie," said the Brown–eyed Boy, breaking the shell and giving it to him again; "even people eat snails, I´ve heard."

于是，有一天，考比飞过玉米地，飞过树梢，来到了小河边。他的翅膀悄无声息，所以，他飞过来的时候，那个“棕眼睛男孩”和“蓝眼睛女孩”没有听见。当他落在附近的一块小岩石上，突然“呱”的一声大叫起来时，两个孩子都被吓了一跳，就好像一个孩子突然对另一个孩子“嗨”地大喊一声，想让他大吃一惊似的。接着，这只鸟儿唱起了叽叽喳喳的小曲儿，一边唱，一边在河岸附近找了一个浅水滩，在里面扑棱着翅膀洗起澡来。洗完澡，那个“蓝眼睛女孩”拿来一个小水螺递到他面前。他把小水螺衔在嘴里尝了尝，就丢到了地上。对他来说，这跟一颗鹅卵石差不多。“棕眼睛男孩”把小水螺的壳敲碎，又递到他面前，说："考比，我想你会喜欢这个的。我听说，连人都吃水螺呢。"

Corbie took the morsel and swallowed it, and soon was cracking for himself all the snails his comrades gave him. But that was not enough, for their eyes were only the eyes of children and his bright bird eyes could find them twice as fast. So he waded in the river, playing "I spy" with his foster brother and sister, and beating them, too, at the game, though they had hunted snails as many summers as he had minutes.

考比啄起那一小口水螺肉，吞了下去。很快，两个小伙伴捡来的水螺都由他自己弄碎吃了。可是这哪够啊，两个孩子的眼睛毕竟只是孩子的眼睛，他自己那双明亮的眼睛找起水螺来，速度要比他们快两倍呢。于是考比在河里趟来趟去，和收养他的姐弟俩玩起了"视觉大发现"游戏，虽然，姐弟俩捉水螺已经捉了好几个夏天了，而考比只不过才捉了几分钟而已，但是比赛的时候，每次都是他赢，他俩输。

He enjoyed doing many of the same things the

children did. It was that, and his sociable, merry ways, that made him such a good playfellow, and because he wanted them to be happy in his pleasure and to praise his clever tricks. Like other children, eating when he was hungry gave him joy, and at times he made a game of it that was fun for them all. Every now and then he would go off quietly by himself, and fill the hollow of his throat with berries from the bushes near the river-bank and, flying back to his friends, would spill out his fruit, uncrushed, in a little pile beside them while he crooned and chuckled about it. He seemed to have the same sort of good time picking berries in his throat cup and showing how many he had found that the children did in seeing which could first fill a tin cup before they sat down on the rocks to eat them.

考比喜欢做很多孩子们喜欢做的事情，再加上他喜欢热闹，而且总是一副欢欢喜喜的样子，所以，大家都喜欢和他玩，而他则想让大家分享自己的快乐，想让大家赞赏他那些聪明的小伎俩。和其他孩子一样，正饿的时候能有东西吃会让他乐不可支，有时候，他会把吃饭当成玩游戏，结果大家都很开心。时不时地，考比会自个儿悄悄溜出去，到河岸附近的灌木丛里塞满一喉咙浆果，然后飞回到朋友们跟前，把采到的浆果完好无损地吐出来，堆成一个小堆，与此同时，他还哼哼唧唧地低声唱着,咕咕咕地轻声笑着,忙得不亦乐乎。采摘浆果的孩子们,总是先要比比谁的锡杯先装满,然后才会找个石头坐下来吃,那会让他们很开心。考比和那些采摘浆果的孩子们一样，当他把摘来的浆果装到他的“喉杯”里拿给大家看时,他也很开心呀!

One day the Brown-eyed Boy and the Blue-eyed Girl were down by the river, hunting for pearls. A pearl-hunter had shown them how to open freshwater clamshells without killing the clams. Suddenly Corbie walked up and, taking one of these hard-shelled animals right out of their hands, he flew high overhead and dropped it down on the rocks near by. Of course that broke the shell and of course Corbie came down and ate the clam, without needing any vinegar or butter on it to make it taste good to him. How he learned to do this, the children never knew. Perhaps he found out by just happening to drop one he was carrying, or perhaps he saw the wild crows drop their clams to break the shells: for after nesting season they used often to come down from the mountainside to fish by the river for snails and clams and crayfish, when they were not helping the farmers by eating up insects in the fields.

一天,“棕眼睛男孩”和“蓝眼睛女孩”正沿河寻觅珍珠。一个采珠人曾经教过他俩怎样打开蚌壳而不伤害里面的蚌。突然,考比走了过来,直接从他们的手里叼走了其中一个带壳的蚌,他飞到他们头顶,飞得高高的,然后把那个硬壳动物丢到了附近的岩石堆上。不用说,蚌壳摔碎了;也不用说,考比飞下来吃掉了蚌肉,也不蘸点醋、抹点黄油弄得好吃一点儿。他是怎么学会这一手的,孩子们从来都不知道。也许偶尔有一次嘴里的蚌不小心掉下去摔碎了,他就发现了?也许那些野乌鸦把嘴里的蚌丢下来摔碎的时候他看到了?因为产卵季节过后,就不用再帮着农民消灭田里的昆虫了,所以,他们就会从山坡上飞到河边捕食蜗牛、河蚌和小龙虾。

Corbie liked the crayfish, too, as well as people like lobsters and crabs, and he had many an exciting hunt, poking under the stones for them and pulling them out with his strong beak.

考比也喜欢吃小龙虾,喜爱的程度绝不亚于人类对大龙虾和螃蟹的喜爱。他已经有过很多次激动人心的捕虾经历了:先在石头下面戳一戳、捅一捅,然后用坚硬的嘴巴把找到的小龙虾拽出来。

There seemed to be no end of things Corbie could do with that beak of his. Sometimes it was a little crowbar for lifting stones or bits of wood when he wanted to see what was underneath; for as every outdoor child, either crow or human, knows, very, very interesting things live in such places. Sometimes it was a spade for digging in the dirt. Sometimes it was a pick for loosening up old wood in the hollow tree where he kept his best treasures. Sometimes it worked like a nut-cracker, sometimes like a pair of forceps, and sometimes—oh, you can think of a dozen tools that beak of Corbie´s was like. He was as well off as if he had a whole carpenter´s chest with him all the time. But mostly it served like a child´s thumb and forefinger, to pick berries, or to untie the bright hair–ribbons of the Blue–eyed Girl or the shoe–laces of the Brown–eyed Boy. And once in a long, long while, when some stupid child or Grown–Up, who did not know how to be civil to a crow, used him roughly, his beak became a weapon with which to pinch and to strike until his enemy was black and blue. For Corbie learned, as every sturdy person must, in some way or other, how to protect himself when there was need.

考比的嘴巴好像无所不能。有时候，他的嘴巴就是一根小撬棍，能让他撬开石头或者木块看看下面到底藏着什么，因为只要是喜欢户外活动的孩子，无论是乌鸦的孩子还是人类的孩子，谁不知道这种地方会住着特别、特别有趣的东西呢？有时候，他的嘴巴就是一把铁锹，他会用它在土里到处乱挖；有时候，他的嘴巴是一把小铁镐，可以把空心树上的老木头弄松，把他最好的珍宝放进去保存起来；有时他的嘴巴像核桃夹，有时像钳子，有时像……哦，他嘴巴像的工具太多了，你能想出一大堆来。总之，他的工具可齐全啦，好像他身上随时都带着木工箱似的。不过，大多数情况下他的嘴巴像是小孩子的拇指和食指，他用它采摘浆果，或者解开“蓝眼睛女孩”那漂亮的发带和“棕眼睛男孩”的鞋带。偶尔，当然这样的情况极其少见，如果哪个愚蠢的孩子或者大人不知如何善待乌鸦，胆敢对他动粗，那他的嘴巴就会变成武器，追着敌人连掐带拧、又咬又啄，直到敌人全身青一块、紫一块才罢休，因为像所有坚强不屈的人那样，考比不知怎么也学会了如何在必要的时候进行自我保护。

Yes, Corbie´s beak was wonderful. Of course, lips are better on people in many ways than beaks would be; but we cannot do one tenth so many things with our mouths as Corbie could with his. To be sure, we do not need to, for we have hands to help us out. If our arms had grown into wings, though, as a bird´s arms do, how should we ever get along in this world?

是呀，考比的嘴巴真是太奇妙了。当然了，在很多方面，人类的嘴唇比鸟喙要优越多了，可是，我们用嘴巴做的事情还不及考比的十分之一呢。的确，我们不需要用嘴巴做那么多事情，因为我们还有一双手呢。可是，假如当初我们的胳膊也像鸟的胳膊一样变成了翅膀，那我们该如何在这个世界上生存呢？

The weeks passed by. A happy time for Corbie, whether he played with the children or slipped off and amused himself, as he had a way of doing now and then, after he grew old enough to feel independent. The world for him was full of adventure and joy. He never once asked, "What can I do now to amuse me?" Never once. His brain was so active that he could fill every place and every hour full to the brim of interest. He had a merry way about him, and a gay chatter that

Corbie slipped off and amused himself.
考比偷偷溜出去自娱自乐了。

seemed to mean, "Oh, life to a crow is joy! JOY! " And because of all this, it was not only the Brown-eyed Boy and the Blue-eyed Girl who loved him. He won the hearts of even the Grown-Ups, who had sometimes found it hard to be patient with him during the first noisy days, when he tired them with his frequent baby "kah-and-gubble," before he could feed himself.

时间一个星期、一个星期地过去了。对考比来说，这是一段快乐的时光，无论和孩子们玩耍，还是偷偷溜出去自娱自乐都是如此，因为自从他长大自立之后，他就有办法时不时地偷偷溜出去自娱自乐了。对他来说，这个世界充满了冒险和欢乐。他从来也没有问过这样的问题："现在我干点什么才能让自己开心呢？"一次也没有。他的大脑太活跃了，无论何时何地他都兴致勃勃、不知倦怠。他是个乐天派，总是喋喋不休、兴高采烈地欢叫着，似乎在说："噢，对乌鸦来说，生活就是欢乐！是欢乐！"因为以上这些原因，他不仅得到了"棕眼睛男孩"和"蓝眼睛女孩"的青睐，甚至还赢得了大人们的心。不过，在最初那段闹哄哄的日子里，那些大人们有时候会觉得很难对他耐心，因为那个时候，他还不会自己吃饭，总是一个劲儿地叫着"嘎——饱——嘎——饱。"

But, however bold and dashing he was during the day, whatever the sunny hours had held of mirth and dancing, whichever path he had trod or flown, whomever he had chummed with—when it was the time of dusk, little Corbie sought the one he loved best of all, the one who had been most gentle with him, and snuggling close to the side of the Blue-eyed Girl, tucked his head into her sleeve or under the hem of her skirt, and crooned his sleepy song which seemed to mean:

可是,无论他白天多么大胆、多么活泼,无论阳光下他有过怎样的欢欣和雀跃,无论他踏上过或飞越过哪条小路,也无论他与谁交了朋友,只要黄昏来临,小考比就去找他最喜欢的那个人了,那个一直待他最温柔的人。考比依偎在"蓝眼睛女孩"的身边,把脑袋塞进她的衣袖里或者裙摆下,睡眼蒙眬地低声哼唱着,好像在说:

Oh! Soft and warm the crow in the nest

巢里的小乌鸦呀,温暖又舒适,

Finds the fluff of his mother's breast.

哦,那是妈妈怀里茸茸的毛。

Oh! Well he sleeps, for she folds him tight—

瞧，他睡得多香，因为有她紧紧的拥抱，

Safe from the owl that flies by night.

黑夜里猫头鹰飞来也不怕。

Oh! Far her wings have fluttered away,

哦，妈妈拍拍翅膀飞走了，

Nor does it matter in the day.

不过这在白天没关系。

But keep me, pray, till again't is light,

可是求你晚上留下来，

Safe from the owl that flies by night.

黑夜里猫头鹰飞来也不怕。

Thus, long after he would have been weaned, for his own good, from such care, had he remained wild, Corbie, the tame crow, claimed protection with cunning, cuddling ways that taught the Blue-eyed Girl and her brother and the Grown-Ups, too, something about crows that many people never even guess. For all their rollicking care-free ways, there is, hidden beneath their black feathers, an affection very tender and lasting; and when they are given the friendship of humans, they find touching ways of showing how deep their trust can be.

因此，假如当初考比一直待在野外，没有被收养，那他早就得不到这样悉心的呵护了，那对他是有好处的。可是，你瞧，现在这个家养的乌鸦，却还在黏黏糊糊、哼哼唧唧地闹着要别人的保护，这让“蓝眼睛女孩”和她的哥哥，还有那些大人们对乌鸦有了新的了解，而这种了解是许多人永远也无从知晓的。尽管他们嬉笑玩耍，总是一副无忧无虑的样子，可是在它们那黑色羽毛下面，却蕴藏着一颗非常敏感稚嫩的心，那颗心里洋溢着一种永久不变的爱。一旦得到了人类的友谊，它们就会以各种令人感动的方式向人类表达它们对人类的信赖有多深。

Before the summer was over, Corbie had as famous a collection as his great grandfather. The children knew where he kept it, and used sometimes to climb up to look at his playthings. They never disturbed them except to take out the knitting-needle, thimble, spoons, or things like that, which were needed in the house. The bright penny someone had given him, the shiny nails, the brass-headed tacks, the big white feather, the yellow marble, all the bits of colored glass, and an old watch, they left where he put them; for they thought that he loved his things, or he would not have hidden them together; and they thought, and so do I, that he had as much right to his treasures to look at and care for as the Brown-eyed Boy had to his collection of pretty stones and the Blue-eyed Girl to the flowers in her wild garden.

夏天还没结束，考比收藏的宝贝就已经和他爷爷的爷爷当初收藏的宝贝一样闻名遐迩了。孩子们都知道他把宝贝藏在哪里，有时他们会爬到树上去看考比的宝贝玩具。他们从不乱动乱拿，不过有时会把毛线针、顶针、汤匙，或者诸如此类的东西拿走，因为这些东西家里还等着用呢。别人给他的那枚亮晶晶的硬币，那些磨得光光的钉子和铜头大头针，还有那根大大的白羽毛，那颗黄灿灿的大理石，以及各式各样的彩色玻璃碎片和一块旧手表，这些东西他们都没动。他们想考比一定非常喜欢这些东西，要不他就不会把它们堆在一起藏起来。他们想，我也是这么想的，考比有权利拥有并观赏把玩他的这些宝贝，正如“棕眼睛男孩”有权收集那些漂亮的石头，“蓝眼睛女孩”有权拥有野生园里的那些花儿一样。

After his feathers were grown, in the spring, Corbie had been really good-looking in his black suit; but by the first of September he was homely again. His little side-feather moustache dropped out at the top of his beak, so that his nostrils were uncovered as they had been when he was very young. The back of his head was nearly bald, and his neck and breast were ragged and tattered.

春天，考比羽翼丰满，全身上下一身黑色的套装，看上去可真是漂亮呀！可是一到九月就又像以前那样不好看了。嘴巴上面像胡子一样的侧羽全掉光了，结果就像他小时候那样两个鼻孔全都露出来了。后脑勺几乎全秃了，脖颈和胸前的羽毛也破破烂烂、褴褛不堪。

Yes, Corbie was molting, and he had a very unfinished sort of look while the new crop of paint-brushes sprouted out all over him. But it was worth the discomforts of the molt to have the new feather coat, all shiny black; and Corbie was even handsomer than he had been during the summer, when cold days came, and he needed his warm thick suit.

是的，考比正在换羽，此时的他，全身上下冒出了一层像油彩笔那样的新羽毛，看上去很有点不修边幅的样子。不过，为了穿上一件乌黑闪亮的新羽衣，换羽的种种不适还是值得的。天气变冷的时候，换上新衣的考比甚至比夏天时还要漂亮，再说，那个时候他也需要一件又厚又暖和的新衣服。

At this time all the wild crows that had nested in that part of the country flew every night from far and wide to the famous crow-roost, not far from a big peach orchard. They came down from the mountain that showed like a long blue ridge against the sky. They flew across a road that looked, on account of the color of the dirt, like a pinkish-red ribbon stretching off and away. They left the river-edge and the fields. Every night they gathered together, a thousand or more of them. Corbie's father and mother were among them, and Corbie's two brothers and two sisters. But Corbie was not with those thousand crows.

就在考比换羽的时候，每天晚上，在那个地方筑巢的野乌鸦都纷纷从四面八方向著名的乌鸦栖息地飞去，那里不远处有一个大大的桃园。他们从一座山上飞下来，那山映着天空就像是一条长长的蓝色脊线；他们穿过一条大道，那条大道上铺满红红的泥土，就像是一条飘向远方的粉红丝带；他们离开了河滩、离开了田野。

每天晚上他们都聚集在一起，一千只或者比这还多。考比的爸爸妈妈也在其中，还有考比的两个兄弟和两个姐妹，可是，这上千只乌鸦里面却没有考比的身影。

No cage held him, and no one prevented his flying whither he wished; but Corbie stayed with the folk who had adopted him. A thousand wild crows might come and go, calling in their flight, but Corbie, though free, chose for his comrades the Brown-eyed Boy and the Blue-eyed Girl.

没有笼子囚禁他，没人拦着他，他想飞到哪儿就飞到哪儿。可是，考比还是留在了收养他的人们身边。一千只野乌鸦飞来了，又飞走了，他们飞着，召唤着，可是，考比，尽管他是自由的，但是他还是选择了他亲密的伙伴——那个“棕眼睛男孩”和“蓝眼睛女孩”。

I thought all along it would be so if they were good to him; and that is why I said, the day he was kidnapped, that you need not be sorry for Corbie—not very.

我一直以为只要他们对考比好，事情的结果就会是这样的。这就是为什么，那天考比被绑架的时候，我说你不必为他难过——不必很难过。

8

Ardea's Soldier
白鹭阿尔蒂巢营里的士兵

In years long gone by, soldiers called "knights" used to protect the rights of other people; and, when the weak were in danger, these soldiers went forth to fight for them. They were so brave, these knights of old, that there was nothing that could make them afraid. Dragons even, which looked like crocodiles, with leather wings and terrible snatching claws and fiery eyes and breath that smoked–dragons, even, so the stories go, could not turn a knight away from his path of duty. Mind, I am not telling you that there ever were creatures that looked like that; but certain it is that there were dangers dreadful to meet, and "dragon" is a very good name to call them by.

很多年以前，我们把士兵称作“骑士”，是他们捍卫着他人的权利。当弱者遇到危险的时候，那些士兵就挺身而出为他们而战。这些古时候的骑士，他们非常勇敢，任何事情都不会吓倒他们。即使那种酷似鳄鱼的恶龙，那种长着皮翅膀、脚爪锋利可怕、眼露凶光、口中吐火的恶龙，故事里是这样描述的，即便是那样的恶龙都无法让骑士们望而生畏，逃离自己的职责。注意，我可不是说曾经真有过这种可怕的龙，可是有一件事情是肯定的，那就是过去的确曾经有过种种可怕的危险，而“龙”就是这些危险的代名词。

You know, do you not, that there are soldiers, still, who protect the rights of others; and although we do not commonly call them "knights," they still fight for the weak, and are so brave that dangers as fearsome as dragons, even, cannot scare them.

你知道的，难道你不知道？现在仍然有士兵捍卫着他人的权利。虽然我们一般不把他们称作“骑士”，但他们仍在为弱者而战，而且英勇无畏，就算像“龙”一样可怕的危险也吓不倒他们。

There was such a soldier in Ardea´s camp; and if he had lived in olden days, he would probably have been called "Knight of the Snowy Heron."

阿尔蒂的巢营里就有这样一位士兵。如果他生活在很久以前，很有可能被称为“白鹭骑士”。

Ardea was a bride that spring, and perhaps never was there one much lovelier. Her wedding garment was the purest white; and instead of a veil she wore, draped from her shoulders, snowy plumes of rare beauty, which reached to the bottom of her gown, where the dainty tips curled up a bit, then hung like the finest fringe.

那年春天，阿尔蒂做了新娘，世上也许没有比她可爱的新娘了。她穿着世界上最纯净最洁白的婚纱，虽然没有细纱遮面，但是却披了一身美丽的羽衣，雪白雪白的，举世罕见，从肩膀一直拖到婚纱的底摆，底摆处那小巧精致的羽毛尖微微翘起，然后像穗一样垂在那里，世上再也找不出这样漂亮精美的穗边了。

The Soldier watched her as she stood alone at the

edge of the water, so small and white and slender against the great cypress trees bearded with Spanish moss, and thought she made a picture he could never forget. And when her mate came out to her, in a white wedding-robe like her own, with its filmy cape of mist-fine plumes, Ardea's Soldier smiled gently, for he loved Heron Camp and shared, in his heart, the joys of their home-coming.

阿尔蒂独自伫立在水边,"士兵"远远地注视着她:高高的柏树,柏树上长满了西班牙苔藓,像是密密麻麻的胡子,她站在树下,看上去如此娇小、纤细、洁白无瑕,她觉得那就是一幅画,一幅她永远也不会被忘记的画。她的丈夫向她走来,披着一件薄如蝉翼的婚礼斗篷,和她的婚纱一样洁白,上面的羽毛像薄雾一般漂亮。阿尔蒂巢营里的"士兵"轻轻地笑了,他非常喜欢"鹭营",他在心里默默地分享着他们回家的喜悦。

Ardea and her mate took a pleasant trip, looking for a building place at the edge of a swamp. They did not object to neighbors; which was fortunate, as there were so many other herons in the camp that it would have been hard to find a very secret spot for their nest. After looking it over and talking about it a bit, they

She wore, draped from her shoulders, snowy plumes of rare beauty.
她身披美丽的羽衣,雪白雪白的,举世罕见。

chose a mangrove bush for their very own. They had never built a house before, but they wasted no time in hunting for a carpenter or teacher, but went to work with a will, just as if they knew how. It was like playing a game of "five-six, pick up sticks"; only they did not lay them straight but in a scraggly criss-cross sort of platform, with big twigs twelve inches long at the bottom and smaller ones on top. Then, when it looked all ready for a nice soft lining, Ardea laid an egg right on the rough sticks. Rather lazy and shiftless, don´t you think? or maybe they didn´t know any better, poor young things who had never had a home before! Ah, but there was another pair of snowy herons building in the bush next door, and they didn´t put in anything soft for their eggs, either; and six or eight bushes farther on, a little blue heron was already sitting on her blue eggs in almost exactly the same sort of nest.

阿尔蒂和她的丈夫去沼泽边寻找筑巢的地方,这是一次愉快的旅行。他们并不排斥邻居，这真是万幸,因为鹭营里各种各样的鹭鸟很多,很难找到一个非常隐秘的地方筑巢。经过一番仔细观察和认真商量,他们决定在一片红树林里安家。以前他们从未建过房子,但他们没有浪费时间四处去找木匠或去请老师,而是立刻投入了工作,非常执着,似乎他们知道怎么建房子一样。他们修房子就像是在玩“四、五、六，拿起棍”的游戏，只不过他们不是把树枝整整齐齐地搭在一起，而是稀稀拉拉、横七竖八地堆成一个小平台，十二英寸长的大树枝铺在下面，小的则铺在上面。然后，就在一切准备就绪、可以开始铺设柔软舒适软垫的时候，阿尔蒂在粗糙不平的小棍上产了一枚蛋。她也太懒了点吧，也太能凑合了点吧？要不就是他们考虑得不够周到，可怜他们以前从来没有过家呀！是呀，不过，隔壁的灌木丛里，还有一对白鹭也在筑巢，他们也没有为自己的蛋铺什么软垫子；再往那边隔了六七个灌木丛的地方，一只小蓝鹭已经坐在她那蓝颜色的鹭蛋上开始孵蛋了，她的窝和白鹭的几乎一模一样。

So that is the kind of carpenters herons are! Sticks laid tangled up in a mass is the way they build! Yes, that is all—just some old dead twigs. I mean that is all you could *see*; but never think for a minute that there wasn´t something else about that nest; for Ardea and her mate had lined it well with love, and so it was, indeed, a home worth building.

瞧，这就是鹭木匠！他们建房子，把小棍子乱七八糟地堆在一起就完事！是的，就这么简单，只不过是些枯老的小树枝罢了。我的意思是说你能“看”见的就这些，可千万不要以为那个巢就没别的什么了，千万可别这么想，哪怕一分钟都别这么想，因为那里面铺的是阿尔蒂和她丈夫的爱，所以说嘛，这个家可是的的确确值得一建啊。

In less than a week there were four eggs beneath the white down comforter that Ardea tucked over them; and the little mother was as well pleased as if she had had five, like her neighbors, the other snowy heron and the little blue heron.

不到一周，阿尔蒂就把四枚蛋掖在了白色的羽绒被里，她的邻居——另外那只白鹭和那只小蓝鹭，她们都有五枚蛋，不过这位小妈妈和她们一样，可高兴了。

If the eggs of the little blue heron were blue, would not those of the snowy herons be pure white? No, the color of eggs does not need to match the color of feathers; and Ardea´s eggs and those of her next–bush neighbor were so much like the beautiful blue ones of the little blue heron, that it would be very hard

Near Ardea's Home.
阿尔蒂家附近。

for you to tell one from the other. Perhaps Ardea could not have told her own eggs if she had not remembered where she had built her nest. As it was, she made no mistake, but snuggled cosily over her pretty eggs, doubling up her long slender black legs and her yellow feet as best she could.

既然小蓝鹭的蛋是蓝的，那么，白鹭的蛋会不会就是纯白的呢？不一定，因为鸟蛋的颜色不一定非要和羽毛的颜色一致。阿尔蒂产的蛋和隔壁灌木丛里的那个邻居产的蛋一样，都是蓝的，特别像小蓝鹭的那几个蓝蛋，所以，很难区分谁是谁的。如果阿尔蒂当初没有记住自己把巢建在哪儿了，也许她就认不出自己产的蛋了。可实际上，她没有弄错。她使劲蜷起她那又细又长的黑腿，收起黄黄的脚，舒舒服服地把那漂亮的蛋拥在了身下。

If she found it hard to sit there day after day, she made no fuss about it; and probably she really wanted to do that more than anything else just then, since the quiet patience of the most active birds is natural to them when they are brooding their unhatched babies. Then, too, there was her beautiful mate for company and help; for when Ardea needed to leave the nest for food and a change, the father-bird kept house as carefully as need be.

日复一日地坐在巢里很辛苦，可是就算她觉着了，她也没大惊小怪。也可能那个时候她最想做的事情就是安安静静地坐在巢里呢。为了孵化未出世的小宝宝，再活泼、再好动的鸟儿都会变得耐心十足，那是她们的天性。再说了，她那漂亮的丈夫还在一旁陪伴着她，帮着她呢。如果阿尔蒂得出去觅食或透透空气，做爸爸的就会尽心尽力地照料一切。

To her next-bush neighbors and the little blue herons Ardea paid no attention, unless, indeed, one of them chanced to come near her own mangrove bush. Then she and her mate would raise the feathers on the top of their heads until they looked rather fierce and bristly, and spread out their filmy capes of dainty plumes in a threatening way. That criss-cross pile of old dead twigs was a dear home after all, being lined, you will remember, with the love of Ardea and her mate; and they both guarded it as well as they were able.

阿尔蒂一点儿也不在意隔壁灌木丛里的那些邻居，

还有那个小蓝鹭，当然了，如果他们有谁不小心跑到了她的红树丛附近，那就另当别论了。那个时候，她和丈夫会竖起头顶的羽毛，张开身上那像薄雾一样精巧秀美的大羽毛斗篷，装出一副恶狠狠、凶巴巴的样子把他吓走。毕竟，那堆横七竖八的枯树枝是他们心爱的家，里面铺的、垫的都是阿尔蒂和丈夫的爱呀！他们俩都竭尽所能、拼尽全力守护着这个家。

At last the quiet brooding days came to an end, and four funny little herons wobbled about in Ardea´s nest. Their long legs and toes stuck out in all directions, and they couldn´t seem to help sprawling around. If there had been string or strands of moss or grass in the nest, they would probably have got all tangled up. As it was, they sometimes nearly spilled out, and saved themselves only by clinging to the firm sticks and twigs. So it would seem that their home was a good sort for the needs of their early life, just as it was; and no doubt a heron´s nest for a heron is as suitable a building as an oriole´s is for an oriole.

平静的孵卵期终于结束了，阿尔蒂的巢里出现了四只摇摇晃晃、有趣可爱的小白鹭。他们的长腿和脚趾伸得到处都是，好像他们不由自主就要到处乱爬似的。幸亏巢里没有绳呀、藓呀或草呀等乱七八糟的东西，否则这些小白鹭早就被缠得满身都是了。有时候，要不是紧紧地抓住细棍子和小树枝，他们几乎就从巢里跌下去了。如此看来，他们的家的确不错，正适合他们的早年生活，的确如此。所以，金莺肯定有适合金莺的巢，鹭鸟肯定也有适合鹭鸟的巢。

It would take some time before the babies of Ardea would be able to straighten up on their long, slim legs and go wading. Until that day came, their father and mother would have to feed them well and often. Now the marsh where the snowy herons went fishing, where the shallow water was a favorite swimming–place for little fishes, was ten miles or more from their nest. Some kinds of herons, perhaps most kinds, are quiet and stately when they hunt, standing still and waiting for their game to come to them, or moving very slowly and carefully. But Ardea and the other snowy herons ran about in a lively way, spying out the little fishes with their bright yellow eyes, and catching them up quickly in their black beaks.

阿尔蒂的宝宝们要等到能把又长又细的腿伸直、稳稳地站起来的时候，才可以去泥里、水里蹚着玩，可那

That criss-cross pile of old dead twigs was a dear home, and they both guarded it.
那堆横七竖八的老枯树枝是他们心爱的家,他们守护着它。

是需要一段时间的。在此之前,他们的爸爸妈妈得经常给他们喂食,把他们喂得饱饱的。白鹭喜欢去沼泽地里捕鱼,那里的浅水滩是小鱼儿们最喜欢的游泳池,可是,那块沼泽地离他们的巢至少还有十英里呢。有些鹭,也许是大多数鹭,捕食的时候很安静、很端庄,它们一动不动地站在那里,等待着猎物的到来;即使动,也是慢条斯理、轻手轻脚地。可是阿尔蒂和别的白鹭捕食时,却总是瞪着明亮的黄眼睛,快活地跑来跑去,一发现小鱼就飞快地伸出黑黑的嘴巴叼住。

After swallowing a supply of food, Ardea took wing and returned across the miles to her young. Standing on the edge of her nest and reaching down with her long neck, she took the bill of one of her babies in her own mouth, and dropped part of what she had swallowed out of her big throat down into his small one. When she had fed her babies and preened her pretty feathers a bit, she was off again on the ten-mile flight; for many a long journey she and her mate must take ere their little ones could feed themselves. But ten miles over and over and over again were as nothing to the love she had for her children; and faithfully as she had brooded her eggs, she now began the task of providing their meals. She seemed so happy each time she returned, that perhaps she was a little bit worried while she was away; but there is no reason to think she really was afraid that any great harm could come to them.

吞下足够的食物后，阿尔蒂就展翅起飞了，越过数十英里回到孩子们身边。她站在巢边，低头伸出长长的脖子，张开嘴，含住宝宝的嘴，把刚才吞下去的鱼从她那大大的喉咙里吐出一部分来，送到宝宝的小喉咙里。喂完所有的宝宝后，她稍稍梳理一下漂亮的羽毛，就又出发了，踏上了新一轮的十英里之旅。在孩子们还不能自食其力之前，阿尔蒂和她的丈夫必须这样一次又一次地在这段长长的路上往返。可是，和阿尔蒂对孩子们的爱比起来，一次又一次不断往返的十英里根本算不了什么。像当初孵蛋时那样，她尽职尽责地履行着为孩子们提供食物的任务。每次回来，她看上去都是那么快乐，也许是离开孩子们的时候，她有过一点点担心吧？说不上是害怕，因为她怎么会想到孩子们会遭殃呢？

Certainly she was unprepared for what she found when she flew back from her fourth fishing trip. Even when she reached Heron Camp, she did not understand. There are some things it is not given the mind of a bird to know. She could not know, poor dear, that there were people in the world who coveted her beautiful wedding plumes. Women there were, who wished to make themselves look better by wearing the feathers that Nature had given snowy herons for their very own. And men there were, who thought to make themselves grander in the dress of their organization by walking about with heron plumes waving on their heads. The two kinds of white herons with wonderful plumes that have been put to such uses are called Egrets and Snowy Egrets, and the feathers, when they are stripped from the birds, are called by the French name of *aigrette*.

的确,当她完成第四次捕鱼旅行飞回家时,她所看到的事情是她始料未及的。她到了巢营也没明白这是怎么一回事。有些事情可不是鸟儿能理解的。可怜的宝贝,她怎么会知道这个世界上竟会有人垂涎她那美丽的婚羽呢?还有那些女人,为了把自己打扮得更漂亮,竟然把大自然送给白鹭的婚羽配在了自己的婚纱上;还有那些身穿制服的男人,为了显得更加显赫、派头十足,竟然头戴鹭羽招摇过市。有两种白鹭的羽毛非常美丽,经常被用作装饰羽,这两种鹭分别叫"Egret"和"Snowy Egret"(注:都是白鹭的意思),从这些鸟儿身上拔下来的羽毛还有一个法语名字，叫“鹭羽”。

Now, of course, Ardea could not know about this, or that the Plume–Hunters had come to steal her wedding feathers. But she knew well enough that danger was at hand, and that in times of trouble a mother´s place is beside her babies. Her heart beat quickly with a new terror, but she stayed, the brave bird stayed! And all about her the other herons stayed also. They had no way to fight for their lives, and they might have flown far and safely on their strong wings; but none of them would desert the home built with love while the frightened babies were calling to their fathers and mothers.

当然了，这些阿尔蒂是无法知道的，她也无法知道那些“猎羽者”来这里是为了偷她的婚羽。可是，她很清楚危险就在眼前，她也很清楚，危险来临时，母亲应该待在孩子身边。恐惧让她的心怦怦乱跳，但她留了下来，这只勇敢的鸟儿她留了下来！而且，周围其他的白鹭也都留了下来。他们手无寸铁，无法为生命而战，他们完全可以乘着强劲有力的翅膀飞到很远、很安全的地方去。可他们谁也没有弃家而去，那可是他们用爱营造的家呀！那里还有受到惊吓，正使劲呼唤着爸爸妈妈的宝宝们呢。

No, *they* could not fight for their lives, but there was one who could. For danger did not come to Heron Camp without finding Ardea´s Soldier at his post.

是的，"他们"无法为生命而战，可是，有人却能。因为，每次鹭营遇到危险，他都会坚守在自己的岗位上，他就是那个守护阿尔蒂营巢的士兵。

Now the Plume–Hunters did not have bodies like crocodiles and leather wings, you know; but they were dragons of a sort, for all that, for they carried brutal things in their hands that belched forth smoke and pain and death, and they were cruel of heart, and they had sold themselves to do evil for the sake of the dollars that covetous men and women would pay them for feathers.

哦，你知道的，那些"猎羽者"没有恶龙那样的身体，也没有恶龙那样的皮翅膀。可是，尽管如此，他们都是恶龙，因为他们手里拿着的东西凶残无比，能喷出烟雾，带来痛苦、带来死亡；他们内心残酷，为了得到贪图羽毛的那些男男女女的几个臭钱，他们早已把灵魂出卖给了魔鬼。

Dragons though they were, Ardea´s Soldier met them bravely. I like to think how brave he was; for was not the fight he fought a fight for our good old Mother Earth, that she might not lose those beautiful children of hers? If the world should be robbed of Snowy Herons, it would be just so much less lovely, just so much less wonderful. And have they no right to life, since the same Power that gave life to men gave life to them? And when we think about it this way, who seems to have the better right to those plumes—herons, or men and women?

可是，哪怕是凶残的恶龙，守护阿尔蒂营地的士兵也毫无畏惧。我愿意展开想象的翅膀，去想象他是多么的英勇无畏，因为，他是在为我们亲爱的大地母亲而战，为的是不让她失去那些美丽的孩子；假如没有了白鹭，这个世界就不那么可爱，不那么美丽了；既然赋予人类生命的那种“力量”也同样把生命赋予了他们，他们就有生存的权利。如果这样想的话，我们就不禁会问：谁更有权利拥有这些羽毛呢？是白鹭？还是那些男男女女？

The Soldier believed in Ardea´s right to life, believed in it so deeply that he stood alone before the Plume–Hunters and told them that, while he lived, the birds of his camp should also live.

那个“士兵”坚信阿尔蒂有生存的权利，正是因为他对此坚信不疑，所以，他才敢于独自站在那些“猎羽者”面前，警告他们说：只要我活着，我营地里的鸟儿就不会死。

And that is why they killed him—the dragons who were cruel of heart and had sold themselves to do evil for the sake of dollars that covetous men and women would pay for feathers.

于是，他们杀死了他，这些内心残酷、为了那些贪图羽毛的男男女女的几个臭钱就把灵魂出卖给了魔鬼的恶龙！

Because of his courage and because of the cause for which he died, I think, don´t you, that Ardea´s Soldier might well be called "Knight of the Snowy Heron."

守护阿尔蒂的士兵如此英勇无畏,他为了美好的事业而献出了生命,所以,我们应该授予他“白鹭骑士”的称号,你认为如何呢?

I said that he was alone, and it is true that no one was there at the camp to help him. But many there were in other places doing their bit in the same good fight. Another soldier, named Theodore Roosevelt, did much for these birds when he was President, by granting them land where no man had a right to touch them; for it makes a true soldier angry when the weak are oppressed, and he said, "It is a disgrace to America that we should permit the sale of aigrettes." Another man, named Woodrow Wilson, whose courage also was so great that he always did what he believed to be right, would not permit, when he was Governor of New Jersey, a company to sell aigrettes in that State; he said, "I think New Jersey can get along without blood–money."

我说过，他在孤军奋战。的确，巢营里谁也没有来帮他。不过，其他地方有许多人正在为一场同样正义的战斗而贡献着自己的一份力量。一名叫西奥多·罗斯福的士兵任总统时，为这些鸟儿做了件了不起的事情，他把大量的土地赠给了他们，在那里，任何人都没有权利碰他们。看到弱者受到欺压，一个真正的士兵会愤怒的，所以，他说："允许买卖鹭羽是美国的耻辱！"还有一个人，叫伍德罗·威尔逊，他也英勇无比，只要他认为正确的事情，他做起来都义无反顾。任新泽西州州长时，他禁止任何公司在该州买卖鹭羽。他说："我认为没有这些滴血的钱，新泽西州照样能运转。"

Many another great man, besides, served the cause of Ardea. So many, in fact, that there is not room here to tell about them all. But there is room to say that the children helped. For, you know, every Junior Audubon Society sends money to the National Association of Audubon Societies—not much, but a little; and when the Knight of the Snowy Heron was killed, that little helped the National Association to hire another soldier to take his place. Now, think of that! There was another soldier who so believed in the Herons' right to life and plumage, that he was ready to protect them though it meant certain danger to himself!

另外还有许许多多了不起的人也投身到了保护鹭鸟阿尔蒂的伟大事业中，这样的人太多了，无法在这里一一列举。不过，说说孩子们的贡献还是可以的。你知道，所有“青少年奥杜邦协会”的会员都在给“全国奥杜邦协会”捐款，数目不大，但有一点点。所以，“白鹭骑士”惨遭不测之后，那一点点捐款使得“全国奥杜邦协会”能够重新雇用一个士兵来接替他的岗位。哦，想想吧！又有一位士兵坚信白鹭有活着的权利、有拥有羽毛的权利，而且，他已经做好为这些权利而战的准备，尽管这意味着一定的危险！

Yes, there is to this very day a soldier at Heron Camp. Do you know a way to keep him safe? Why, you children of America can do it if you will, and it need not cost one of you a penny. You can do it with your minds. For if every girl makes up her mind for good and all that she will never wear a feather that costs a bird its life; and if every boy makes up his mind for good and all that he will never be a feather-hunting dragon—why there will not be *anybody* growing up in

America to harm Ardea, will there? You can keep the Soldier of Heron Camp safe by just wishing it! That sounds wonderful as a fairy story come true, does it not? And like the knight in some old fairy tale, could not Ardea´s new Soldier "live happily forever after"?

是的，直到今天，鹭营里总有一名士兵。那你知道怎样保护它吗？对呀，美国的小公民们，只要你们愿意，你们就可以保护它，而且你们谁也不需要花一分钱，只需拿出一颗真诚的心就行了。因为，假如你们所有的女孩都下决心永远不再佩戴用鸟儿生命换来的羽毛，假如你们所有的男孩都下决心永远不再做猎取羽毛的恶魔，那么，等你们长大了，美国就不会再有任何人伤害鹭鸟阿尔蒂了，你们说是不是呢？你们的美好愿望就能保护“鹭营”里的士兵！听起来是不是很奇妙？是不是有点像童话故事变成了现实呢？所以，就像古老的童话故事里的骑士那样，鹭鸟阿尔蒂巢营里新来的士兵“从此过上了幸福的生活”，你们说会不会呢？

9

The Flying Clown
飞翔“小丑”——夜鹰

There are many accounts of the flying clown, in books, nearly all of which refer to him as bull–bat or nighthawk, and a member of the Goatsucker or Night-jar family. But he wasn´t a bull and he wasn´t a bat and he wasn´t a hawk and he wasn´t a jar; and he flew more by day than by night, and he never, never milked a goat in all his life. So for the purposes of this story we may as well give him a name to suit ourselves, and call him Mis Nomer.

书中有很多关于“飞翔小丑”的记述。这些记述几乎无一例外地把他叫作“公牛蝙蝠”或者“夜晚的鹰”，而且说他是“偷吸山羊奶的鸟”或“夜晚发出刺耳声音的罐子”的家族成员。可是他不是公牛，不是蝙蝠，不是鹰，也不是罐子，而且他白天飞得比夜晚多，他一辈子都没有吸过山羊的奶，绝对没有。因此，考虑到讲这个故事的目的，我们不妨把他叫作“错名鸟”先生吧。

He was a poor skinny little thing, but you would not have guessed it to see him; for he always wore a loose fluffy coat, which made him look bigger and plumper than he really was. It was a gray and brown and creamy buff-and-white sort of coat, quite mottled, with a rather plain, nearly black, back. It was trimmed with white, there being a white stripe near the end of the coat-tail, a big, fine, V-shaped white place under his chin that had something the look of a necktie, and a bar of white reaching nearly across the middle of each wing.

“错名鸟先生”是个骨瘦如柴的小可怜，可是，见到他，你绝对不会想到他会那么瘦，因为他总是穿着一件松松垮垮、毛毛乎乎的大衣，让他看上去比实际要高、要胖。大衣的颜色很杂，有的地方是灰的，有的地方是棕的，还有的地方黄白相间，是那种淡淡的黄，不过后背上的颜色比较单一，几乎全是黑的。大衣上洒满了白点，衣摆的末端有一个白色长条，下巴底下的那个地方是一个大大的、漂亮的 V 字形白环，有点像一条领带，两个翅膀上分别横着一条白杠，几乎要穿过翅膀中间了。

These bars would have made you notice his long, pointed wings if he had been near you, and they were well worth noticing; for besides just flying with them,—which was wonderful enough, as he was a talented flier,—he used them in a sort of gymnastic stunt he was fond of performing in the springtime.

如果他离你够近，这些白色的横杠会把你的注意力吸引到他那长长的、尖尖的翅膀上去的，他的翅膀真是值得注意呀，因为他飞得很棒，是一个飞行天才，所以他的翅膀除了用于飞行，还被他用来表演飞行特技，有点像体操绝技，他喜欢在春天表演这种特技。

Perhaps he did it to show off. I do not know. Certainly he had as good a right to be proud of his accomplishments as a turkey or a peacock that spreads its tail, or a boy who walks on his hands. Maybe a better right, for they have solid earth to strut upon and run no risks, while Mis did his whole trick in the air. It was a kind of acrobatic feat, though he had no gymnasium with bars or rings or tight rope, and there was no canvas stretched to catch him if he fell. A circus, with tents, and a gate-keeper to take your ticket, would have been lucky if it could have hired Mis to show his skill for money.

没准他做特技表演是为了炫耀,谁知道呢?的确,他完全有权为自己的成就感到骄傲,正如火鸡或者孔雀有权为自己展开的尾巴感到骄傲,男孩有权为自己能徒手倒立行走感到骄傲一样。也许他的理由更充分,因为毕竟展翅的火鸡、孔雀和徒手倒立行走的小男孩都有坚实的大地做支撑,没有危险可言,而“错名鸟”的整个把戏都是在空中表演的。那简直就是杂技表演,不过他可没有带护栏、支撑圈或者安全绳的体育馆,而且万一掉下来,也没有张开的帆布来接它。那种带帐篷的、有把门人收门票的马戏团如果能雇到“错名鸟”先生,让他为钱献艺,那就太幸运了。

But Mis couldn't be hired. Not he! He was a

free, wild clown, performing only under Mother Nature's tent of wide-arched sky. If you wanted to see him, you could—ticket or no ticket. That was nothing to him; for Mis, the wild clown of the air, had no thought either of money or fame among people.

可是“错名鸟”先生不会受雇于任何人的，雇谁也雇不到他！他是个自由自在、随心所欲的小丑，只在大自然母亲的苍穹下展示技艺。他的表演你想看就能看，有没有票都无所谓，因为门票对他毫无意义。“错名鸟”先生，这个随心所欲的空中小丑，他对人类的钱或者名望毫无概念。

Far, far up, he flew, hither and yon, in a matter-of-fact-enough way; and then of a sudden, with wings half-closed, he dropped toward the earth. Could he stop such speed, or must he strike and kill himself in his fall? Down, down he plunged; and then, at last, he made a sound as if he groaned a loud, deep "boom".

他能飞很高很高，一会儿飞到这里，一会儿飞到那里，一副漫不经心的样子。突然，他两翼半合，径直跌向地面。这么快的速度，他能刹得住吗？是不是他一定得掉到地上摔死了？他一路跌了下来，越来越低……最后他“哄”地大叫了一声，声音沉沉的，像是痛苦的呻吟。

But just at the moment of this sound he was turning, and then, the first anyone knew, he was flying up gayly, quite gayly. Then it wasn´t a groan of fear? Mis afraid! Why the rascal had but to move his wings this way and that, and go up instead of down. He might be within a second of dashing himself to death against the ground, but so sure were his wings and so strong his muscles, that a second was time and to spare for him to stop and turn and rise again toward the safe height from which he dived. A fine trick that! The fun of the plunge, and then the quick jerk at the end that sent the wind groaning against and between the feathers of his wings, with a "boom" loud and sudden enough to startle anyone within hearing.

可是声音还未落地，他就转身了。接下来，谁也没想到，他竟然兴高采烈地飞上去了，他的确十分高兴。这么说，那声大叫不是恐惧的呻吟？“错名鸟”先生会恐惧？简直是笑话！哎，这个小淘气只要稍稍转动一下翅膀，就能飞上去，而不是掉下来。他很可能在一秒钟内冲到地上撞死，不过，他的翅膀挥洒自如，他的肌肉发达有力，一秒钟的时间足以让他腾出空来，“刹车”、转

The Flying Clown
飞翔的小丑

身、起飞，重新飞回俯冲前的安全高度。多么精湛的特技表演！俯冲的乐趣：最后那突然的转身，让他的翅膀呼呼生风；“轰”的一声突如其来的大叫，足以让听见这个声音的人吓一大跳。

Yes, you might have seen the little clown at his tricks without a ticket at the wild–circus gate, for all he cared or knew. What did the children of men matter to him? Had not his fathers and grandfathers and great–grandfathers given high–air circus performances of a springtime, in the days when bison and passenger pigeons inherited their full share of the earth, before our fathers and grandfathers and great–grandfathers had even seen America?

是呀，也许你没买票就看了他的野外马戏表演，对他来说这无所谓，或者他根本就不知道。人类的子孙算什么呀，早在野牛和候鸽来到这里尽情享用他们的世袭领地时，他的爸爸、爸爸的爸爸，还有爸爸的爸爸的爸爸就已经在表演春季高空杂耍了。那个时候，我们的爸爸、爸爸的爸爸、爸爸的爸爸的爸爸还没发现美洲大陆呢。

Was it, then, just for the joy of the season that he played in the air, or was there, after all, someone besides himself to be pleased with the sport? Who knows whether the little acrobat was showing his mate what a splendid fellow he was, how strong of wing and skillful in the tricks of flight? Be that as it may, the mate of Mis was satisfied in some way or other, and went with him on a voyage of discovery one afternoon, when the sky was nicely cloudy and the light pleasantly dull.

那么他在空中的杂耍仅仅是应季的欢娱吗？要不就是他想取悦的人除了他自己还另有其人？谁知道这个小巧可爱的杂技演员是不是在向它的意中人炫耀：瞧，我多棒！看，我飞的特技，翅膀多有劲！动作多娴熟！就算是吧，“错名鸟”先生的意中人还是比较满意的。一天下午，她和“错名鸟”先生一起踏上了探索之旅，那天下午，天阴阴的很舒适，阳光暗暗的很惬意。

Now, like all good parents, Mis and his mate were a bit particular about what sort of neighborhood they should choose for their home; for the bringing up of a family, even if it is a small one, is most important.

现在，和所有的好爸爸好妈妈一样，“错名鸟”先生和他的伴侣对在何处安家开始有点挑剔了，因为养育子女是头等大事，哪怕只有不多几个子女。

A peaceful place and a sunny exposure they must have; there must be good hunting near at hand; and one more thing, too, was necessary. Now, the house–lot they finally decided upon met all four of these needs, though it sounds like a joke to tell you where it was. But then, when a clown goes merrily forth to find him a home, we must not be surprised if he is funny about it. It was where the sun could shine upon it; though how Mis and his mate knew that, all on a dull, dark afternoon, I´m sure I can´t tell. Maybe because there wasn´t a tree in sight. And as for peace, it was as undisturbed as a deserted island. It was, in fact, a sort of island in a sea of air, and at certain times of the day and night there was game enough in this sea to satisfy even such hunters as they.

他们的家环境安静，阳光充足，捕食方便，另外还有一点也很必要。这四点要求他们最终选定的宅基地都达到了，不过如果告诉你那块地方在哪里，你会觉得我是在开玩笑。可是，“小丑演员”高高兴兴地飞去给自己找家的时候，如果他的样子十分滑稽，我们着实不必大惊小怪。他找的那个地方阳光充足，不过，那天下午光线昏暗，黑乎乎的，“错名鸟”先生和他的伴侣究竟是怎么知道的，我确实说不上来，或许是因为那里没有树吧；至于安静的环境，那个地方安静得和荒岛一样；其实，那个地方本身就是天空这个茫茫大海上的一个荒岛，无论是白天还是黑夜，在某些特定的时间，这片大海里的猎物多得连他们这样的猎手都心满意足。

Perhaps they chuckled cosily together when they decided to take their peace and sunshine on the flat roof of a very high building in a very large city. Their house-lot was covered with pebbles, and it suited them exactly. So well that they moved in, just as it was.

他们找到了一个很大很大的城市的一个很高很高的楼顶，决定在那里享受他们的宁静和阳光，也许，在做出决定的那一刻他们舒心地笑了。他们的宅基地铺着鹅卵石，很适合他们，太适合了，所以，他们直接就搬了进去。

Yes, those two ridiculous birds set up housekeeping without any house. Mother Nomer just settled herself on the bare pebbles in a satisfied way, and that was all there was to it. Not a stick or a wisp of hay or a feather to mark the place! And as she sat there quietly, a queer thing happened. She disappeared from sight. As long as she didn´t move, she couldn´t be seen. Her dappled feathers didn´t look like a bird. They looked like the light and dark of the pebbles of the flat roof. Ah, so *that* was the one thing more that was necessary for her home, besides sunshine and peace and good hunting. It must be where she could sit and not show; where she could hide by just looking like what was near her, like a sand-colored grasshopper on the sand in the sun, or a walking-stick on a twig or a butterfly on the bark of a tree.

是的，这两只可笑的鸟儿没有盖房子就开始居家过日子了。“错名鸟妈妈”直接在光秃秃的鹅卵石上安顿下来，一副心满意足的样子，如此而已。连一根做标记的小棍，或者干草，或者羽毛都没有！她静静地坐在那里，突然一件奇怪的事情发生了。她不见了。只要她不动，就看不到她。她身上那斑驳的羽毛看上去不像鸟，倒像是楼顶平台上深浅相间的鹅卵石。哦，原来这就是那很有必要的另外一点，除阳光充足、环境安静、捕食方便之外的第四个选址条件。他们选择的地方必须符合这第四个条件：她卧在那里要不显眼，藏在那里能和周围的环境融为一体，就像那个顶着烈日躲在沙地上的土不拉叽的蚱蜢欢欢，或者那个躲在树枝上、外号叫“会走路的棍子”的假死虫安妮，或者那个躲在树皮上的蝴蝶帆儿。

Yes, Mis´s mate knew, in some natural wise way of her own, the secret of making use of what we call her "protective coloration." This is one of the very most important secrets Mother Nature has given her children, and many use it—not birds alone, but beasts and insects also. They use it in their own wild way and think nothing about it. We say that it is their instinct that leads them to choose places where they cannot easily be seen.

是的，“错名鸟”先生的太太知道如何利用她身上被我们称之为“保护色”的东西，这种聪明才智与生俱来，是大自然妈妈赋予她的孩子们的最重要的生存秘诀，许多孩子都在利用这些秘诀，不仅鸟儿在用，畜儿和虫儿也在用。他们的利用方式各自不同、随心所欲，所以我们说他们选择不易暴露的栖息地完全是出于本能。

If you do not understand exactly what instinct is,

do not feel worried, for there are some things about that secret of Mother Nature that even the wisest men in the world have not explained. But this we do know, that when her instincts led Mother Nomer to choose the pebbly roof as a background for her mottled feathers, she did just naturally very much the same thing that the soldiers in the world-war did when they made use of great guns painted to look like things they were not, and ships painted to look like the waves beneath them and the clouds in the sky above.

如果你不明白究竟什么是本能，也用不着担心，因为关于大自然母亲的这个秘密，有些方面连世上最聪明的人也没有做出过清楚的解释。但是，有一点我们的确知道，那就是当“错名鸟”妈妈出于本能选择了铺满鹅卵石的楼顶作她那斑驳羽毛的背景时，她只是自然而然地去做。同样的事，人类在战争中也会做：士兵们给大型枪炮涂上颜色，遮掩其真实面目；给战舰涂上颜色，让它们看上去像海面的波涛和天空中的云。

Only, the soldiers did not use their protective coloration naturally and by instinct. They did this by taking thought; and very proud they felt, too, of being able to do this by hard study. They talked about it a great deal and the French taught the world a new word, *camouflage*, to call it by. And their war–time camouflage *was* wonderful, even though it was only a clumsy imitation of what Mother Nature did when the feathers of Mother Nomer were made to grow dappled like little blotches of light and dark; or, to put it the other way about, when the bird was led, by her instinct, to choose for the nesting–time a place where she did not show.

只不过，士兵使用保护色不是自然而然出于本能，而是借鉴的结果，他们通过刻苦学习掌握了这个秘密，为此他们分外自豪。这个秘密为他们所津津乐道，法国人还发明了一个新词来称呼这个秘密，这个词就是："伪装"。战场上士兵们的伪装确实奇妙，虽然那只是对大自然妈妈的笨拙模仿。当她让"错名鸟"妈妈的羽毛上长满斑驳的小污点时，或者倒过来说吧，当鸟儿出于本能选择了不易暴露的筑巢场所时，她用的就是伪装术。

Of course, it was not just the gravel on the flat roof that would match her feathers; for there isn´t a house in the land that is nearly so old as one thousand years, and birds of this sort have been building much longer than that. No, so far as color went, Mother Nomer might have chosen a spot in an open field, where there were little broken sticks or stones to give it a mottled look—such a place, indeed, as her ancestors used to find for their nesting in the old days when there were no houses. Such a place, too, as most of this kind of bird still seek; for not all of them, by any means, are roof-dwellers in cities.

当然了，能和她羽毛相配的不是只有楼顶平台上的鹅卵石，因为地上可没有老到将近一千年的房子，而这种鸟早在一千年之前就在找地方安家了。所以，只要颜色符合要求，"错名鸟"妈妈也可能会在空旷的田野上选一个地方安家。那里有折断的小枝或碎石，所以看上去斑驳陆离适于伪装。其实，很久以前，没有房屋的时候，她的祖先们寻寻觅觅的就是这种地方；如今这些鸟儿朝思暮想的依然是这种地方。总之，这些鸟儿并不都是城市楼顶居民。

Our bird with the dappled feathers, however, sat in one little spot on that large roof for about sixteen days and nights, with time enough off now and then to get food and water, and to exercise her wings. When she was away, Mis came and sat on the same spot. If you had been there to see them come and go, you would have wondered why they cared about that particular spot. It looked like the rest of the sunny roof—just little humps of light and dark. Ah, yes! but two of those little humps of light and dark were not pebbles: they were eggs; and if you couldn´t have found them, Mis and his mate could, though I think even they had to remember where they were instead of eye-spying them.

楼顶大大的很宽敞,可是我们这只满身斑驳的鸟妈妈却只占了一块小小的地方,她在那里坐了大约十六个昼夜,偶尔离开去找点吃的和喝的,顺便去练练翅膀,然后马上就回来。她不在的时候,"错名鸟"先生就过来坐到那个地方。假如当时你在场,看到了他们的你来我往,你肯定会奇怪他们为何单单对那个地方情有独钟呢?那个地方和楼顶的其他地方没什么两样,都洒满了金色的阳光,看上去都是一堆一堆斑驳陆离的鹅卵石。是呀,的确没什么两样!可是其中有两堆并非鹅卵石,那是鸟蛋。也许你找不到那两个鸟蛋,可是"错名鸟"夫妇却能,不过我想就算是他俩,找到那两枚鸟蛋靠得也不是眼力,而是记忆。

By the time sixteen days were over, there were no longer eggs beneath the fluffy feathers that had covered them. Instead, there were two little balls of down, though you couldn´t have seen them either, unless you had been about near enough to touch them; for the downy children of Mis were as dappled as his mate and her eggs, and they had, from the moment of their hatching, the instinct for keeping still if danger came near.

等到十六个日日夜夜结束的时候，毛茸茸的羽毛下面已经不是两枚蛋了，而是两个可爱的小绒球，不过，除非你离得很近用手摸摸，否则你也看不到他们，因为“错名鸟”先生的两个绒球似的孩子，和他夫人以及他夫人产下的蛋一样，也长满了圆圆的斑点，而且从破壳而出的那一刻起，他们就具有了面临危险一动不动的本能。

Peaceful enough, indeed, had been the brooding days of Mother Nomer. Something of the noise and bustle, to be sure, of the city streets came up to her; but that was from far below, and things far off are not worth worrying about. Sometimes, too, the sound of voices floated out from the upper windows of the

Peaceful enough, indeed, had been the brooding days.
的确,“错名鸟”妈妈孵蛋时的那些日日夜夜很平静。

building, quite near; but the birds soon became used to that.

的确,“错名鸟”妈妈孵蛋时的那些日日夜夜很平静。当然了，城市街道上的喧嚣也会传到她的耳边,但是，那是在远远的楼下，远处的事情没什么可担心的。有时候，从大楼上面的窗户里飘过来喁喁的说话声,很近,可是,鸟儿们很快就习惯了那声音。

When the twins were but a few days old, however, their mother had a real scare. A man came up to take down some electric wires that had been fastened not far from the spot that was the Nomer home. He tramped heavily about, throwing down his tools here and there, and whistling loudly as he worked. All this frightened little Mother Nomer. There is no doubt about that, for her heart beat more and more quickly. But she didn´t budge. She couldn´t. It was a part of her camouflage trick to sit still in danger. The greater the danger, the stiller to sit! She even kept her eyes nearly shut, until, when the man had cut the last and nearest end of wire and put all his things together in a pile ready to take down, he came to look over the edge of the roof–wall. As he bent to do this, he brushed suddenly against her.

可是，这对双胞胎刚刚出生几天，他们的妈妈却受了一场惊吓。一个男人爬到楼顶来解系在离她家不远的一些电线。他在楼顶走来走去，脚步很重，把工具丢得到处都是，而且一边解电线，一边还大声地吹着口哨。这可吓坏了小“错名鸟”妈妈。肯定吓坏了，因为她的心跳越来越快了。但是，她一动也没动。她不能动！临危不动是她伪装术的一部分。越危险，就越不能动！她几乎连眼睛都闭上了。当那个男人割断最后一根，也是离她最近的一根线头，把他的东西收拾到一起堆成一堆准备拿下去的时候，他又走到楼顶的墙边检查了一遍。就在他弯下腰去检查的时候，突然擦碰了她一下。

Then Mother Nomer sprang into the air; and the man jumped, in such surprise that, had it not been for the wall, he would have fallen from the roof. It would be hard to tell which was the more startled for a moment—man or bird. But Mother Nomer did not fly far. She fell back to the roof some distance from her precious babies and fluttered pitifully about, her wings and tail spread wide and dragging as she moved lamely.

“错名鸟”妈妈一跃而起，飞到了空中。这一惊非同小可，那个男人也跳了起来，如果不是那堵墙，他就从楼顶上掉下去了。那一刻，很难说他们俩谁受的惊吓更大，那个男人？还是那只鸟儿？可是，“错名鸟”妈妈并未飞远。她回到离自己宝贝孩子不远处的楼顶，可怜地扇着翅膀一瘸一拐地走来走去，翅膀和尾巴张得大大的，拖在身后。

She did not look like a part of the pebbly roof now. She showed plainly, for she was moving. She looked like a wounded bird, and the man, thinking he must have hurt her in some way, followed her to pick her up and see what the trouble was. Three times he almost got her. Almost, but not quite. Crippled as she seemed, she could still fumble and flutter just out of reach; and when at last the man had followed her to a corner of the roof far from her young, Mother Nomer sprang up, and spreading her long, pointed wings, took flight, whole and sound as a bird need be.

现在她看上去已经不像楼顶上的鹅卵石了。此时的她清晰可见，因为她在动。她看上去像是一只受了伤的鸟儿，而那个男人以为他一定是把她弄伤了，于是他跟过去想把她捉住看看到底是怎么回事。有三次他几乎要捉住她了。是几乎，不是快要！虽然她一瘸一拐的，可她依然能扑棱一下躲开。最后，她跳到了离她的孩子稍远的楼角，那个男人也跟到了那里。“错名鸟”妈妈一下子跳了起来，张开长长的、尖尖的让她成为一只鸟儿的翅膀，飞走了，安然无恙、毫发未损。

The man understood and laughed. He laughed at himself for being fooled. For it wasn´t the first time a bird had tricked him so. Once, when he was a country boy, a partridge, fluttering as if broken-winged, had led him through the underbrush of the wood-lot; and once a bird by the river-side stumbled on before him, crying piteously, "Pete! Pete! Pete-weet! " and once—Why, yes, he should have remembered that this is the trick of many a mother-bird when danger threatens her young.

那个男人明白了，他大笑起来。他笑自己上当了。这已经不是他第一次上小鸟的当了。有一次在乡下，那时他还是个小孩子，一只好像折断了翅膀的鹌鹑害得他跑遍了小树林里的灌木丛；还有一次在河边，一只鸟儿在他前面磕磕绊绊地跑着，一边跑还一边可怜巴巴地哀叫：“皮特！皮特！皮特——维特！”还有一次……唉！是呀,他本来应该记住的,这其实就是许多鸟妈妈在孩子遇到危险时施展的惯用伎俩。

So he went back, with careful step, to where he had been before. He looked this way and that. There was no nest. He saw no young. The little Nomer twins were not the son and daughter of Mis, the clown, and Mother Nomer, the trick cripple, for nothing! They sat there, the little rascals, right before his eyes, and budged not; they could practice the art of camouflage, too.

所以他蹑手蹑脚地回到了刚才那个地方，这里看看,那里瞧瞧。没有鸟巢,也没看到雏鸟。这对小“错名鸟”双胞胎的这个儿子和女儿可不是白当的！谁让他们的爸爸“错名鸟”先生是小丑演员呢？谁让他们的妈妈“错名鸟”夫人是诡计多端的跛脚婆呢？这两个淘气鬼就坐在他的眼前,一动不动。他们也会展示伪装术！

But as he stood and looked, a wistful light came into the eyes of the man. It had been many years since he had found nesting birds and watched the ways of

The little rascals could practise the art of camouflage.
两个淘气鬼也会展示伪装术。

them. His memory brought old pictures back to him. The crotch in the tree, where the robin had plastered her nest, modeling the mud with her feathered breast; the brook-edge willows, where the blackbirds built; the meadow, with its hidden homes of bobolinks; and the woods where the whip-poor-wills called of nights. His thoughts made a boy of him again, and he forgot everything else in the world in his wish to see the little birds he felt sure must be among the pebbles before him. So he crept about carefully, here and there, and at last came upon the children of Mis. He picked up the fluffy little balls of down and snuggled them gently in his big hands for a moment. Then he put them back to their safe roof, and, gathering up his tools, went on his way, whistling a merry tune remembered from the days when he trudged down Long-ago Lane to the pasture, for his father´s cows. Late of afternoon it used to be, while the nighthawks dashed overhead in their air-hunts, showing the white spots in their wings that looked like holes, and sometimes making him jump as they dropped and turned, with a sudden "boom".

他站在那里看着，一副怅然若失的神情。他已经很多年没有见过筑巢的鸟儿了，也很多年没有观察过它们的生活了。记忆将一些久违了的画面带回了他的脑海中：树杈上，知更鸟在用长满羽毛的前胸抹泥筑巢；溪边的柳树上，乌鸫鸟在修建房屋；草地上，食米鸟的鸟窝掩映其中；树林里，北美夜鹰歌唱着茫茫的暗夜。回忆让他再次童心大发，为了看到它们，他忘记了一切。他确信它们就藏在眼前的鹅卵石中，于是，他趴在地上小心翼翼地四处寻找起来，最后终于发现了“错名鸟”夫妇的孩子。他伸出两只大手，拿起两只软乎乎的小绒球，轻轻地捧在手里看了一会儿。然后，他把它们放回了原处，收齐自己的工具，再次往楼下走去。一边走，一边吹起了轻松愉快的口哨。那首曲子是当年他沿着“很久小径”一路跋涉去草原为父亲放牛时唱过的。那通常是在傍晚时分，那时空中捕食的夜鹰在头顶横冲直撞，翅膀上的白点一览无余，像一个个小洞，它们急速下跌转身的时候，突然发出“轰”的一声，有时候能把他吓一大跳。

No sooner had the sound of his whistle gone from the roof, than Mother Nomer came back to her houseless home—any spot doing as well as another, now that the twins were hatched and able to walk about. As she called her babies to her and tucked them under her feathers, her heart still beating quickly with the excitement of her scare, it would be easy to guess from the dear way of her cuddling that it isn´t a beautiful woven cradle or quaint walls of clay that matter most in the life of young birds, but the loving care that is given them. In this respect the young orioles, swinging in their hammock among the swaying tips of the elm tree, and the children of Eve and Petro, in their wonderful brick mansion, were no better off than the twins of Mis and Mother Nomer.

他的口哨声刚一消失,“错名鸟”妈妈就回到了她那无遮无拦的家里。现在双胞胎已经孵出来了,而且已经行动自如,所以,哪里都能当家。她把孩子们叫到跟前,拥在翅膀下面,此时她还惊魂未定,心怦怦地乱跳着。看着她拥抱孩子们时那亲切慈爱的样子,就不难猜到,在幼鸟的一生中,最重要的不是编织精美的摇篮,也不是精心筑造的泥墙,而是父母给予的关爱。从这点来看,在摇曳的榆树梢上搭起吊床飘来荡去的小黄鹂,还有住在奇妙的砖厦里的小燕子,都不如“错名鸟”夫妇家的双胞胎幸福。

Busy indeed was Mis in the twilights that followed the hatching of his children; and, though he was as much in the air as ever, it was not the fun of frolic and clownish tricks that kept him there. For, besides his own keen appetite, he had now the hunger of the twins to spur him on. Such a hunter as he was in those days! Why, he caught a thousand mosquitos on one trip; and meeting a swarm of flying ants, thought nothing at all of gobbling up five hundred before he stopped. Countless flies went down his throat. And when the big, brown bumping beetles, with hard, shiny wing-covers on their backs and soft, fuzzy velvet underneath, flew out at dusk, twenty or thirty of them, as likely as not, would make a luncheon for Mis the clown. For he was lean and hungry, and he ate and ate and ate; but he never grew fat. He hunted zigzag through the twilight of the evening and the twilight of the dawn. When the nights were bright and game was plenty, he hunted zigzag through the moonlight. When the day was dull and insects were on the wing, he hunted, though it was high noon. And many a midnight rambler going home from the theatre looked up, wondering what made the darting shadows, and saw Mis and his fellows dashing busily above where the night-insects were hovering about the electric lights of the city streets. He hunted long and he hunted well; but so keen was his appetite and so huge the hunger of his twins, that it took the mother, too, to keep the meals provided in the Nomer home.

在孩子们孵出来后接下来的黄昏和黎明，“错名鸟”先生真是忙得不可开交。虽然他还像以前那样频繁地出现在空中，但是，他来不是为了嬉戏玩耍、表演小丑把戏。现在，迫使他来到空中的除了自己旺盛的食欲，还有这对嗷嗷待哺的双胞胎。那个时候，他简直是个不折不扣的“猎人”！瞧！他一次就能捉一千只蚊子；遇到一大群飞蚁，一口气吞掉五百只对他来说也根本不在话下；数不尽的苍蝇也被他吞进了肚子里。傍晚时分，大个头的棕色甲虫跌跌撞撞地飞出来了。它们背上背着硬硬的、闪闪发光的翅鞘，肚子底下是软软的、毛乎乎的丝绒，二十只或三十只就很可能够小丑演员“错名鸟”先生吃一顿了。他精瘦精瘦的，总是饥肠辘辘，所以他不停地吃呀吃呀，可是总也胖不起来。傍晚的夜幕和清晨的晨曦中，他上下翻飞、忽左忽右，辛勤地捕食。月光皎洁的夜晚，猎物丰富，他就趁着月光捕食，一会儿飞到东边，一会儿又飞到西边。日光昏暗的白天，昆虫在飞，即使是正午他也去捕食。午夜，许多从剧院漫步回家的人抬头仰望，不知那匆匆飞来的影子是谁投下的。他们看到了正在城市街灯的上空忙着飞来飞去的“错名鸟”先生和他的伙伴们，原来，街灯的四周有无数的夜虫在盘旋。“错名鸟”先生猎食时间长，效率高，可是他太能吃了，而且他的小双胞胎也总是饥饿难耐，所以，为了不让家里断粮，“错名鸟”妈妈也得上阵。

I think they were never unhappy about it, for there is a certain satisfaction in doing well what we can do; and there is no doubt that these birds were made to be hunters. Mis and his kind swept the air, of course, because they and their young were hungry; but the game they caught, had it gone free to lay its myriad eggs, would have cost many a farmer a fortune in sprays to save his crops, and would have added untold discomfort to dwellers in country and city alike.

我想对此他们从未有过抱怨。毕竟，能把力所能及的事情做好会带来某种满足感。毫无疑问，这些鸟儿就是为捕猎而生的。不用说，“错名鸟”先生和伙伴们在空中来了个大扫荡，因为他们的孩子和他们自己正饿呢。不过，假如他们不把这些猎物捉住，任它们而去，它们就会产很多很多卵，让许多农民花大笔的钱去买农药救他们的庄稼，它们还会给住在乡村和城市的人们增添无尽的烦恼。

Although Mis, under his feathers, was much

smaller than one would think to look at him, there were several large things about him besides his appetite. His mouth was almost huge, and reached way around to the sides of his head under his eyes. It opened up more like the mouth of a frog or a toad than like that of most birds. When he hunted he kept it yawning wide open, so that it made a trap for many an unlucky insect that flew straight in, without ever knowing what happened to it when it disappeared down the great hollow throat, into a stomach so enormous that it hardly seems possible that a bird less than twice the size of Mis could own it.

尽管穿着羽衣的"错名鸟"先生比实际看上去要小得多，但是，他不仅胃口很大，别的也很大。他的嘴巴几乎可以说是巨大无比，一直连到脑袋侧面眼睛的下方，张开时不像大多数鸟的鸟嘴，却更像青蛙嘴或蛤蟆嘴。捕猎的时候，他一直大张着嘴，像个陷阱，结果许多不幸的虫儿直接飞到了他的嘴里，这些虫儿从那个大大的喉管里掉进他的胃里时，根本不知道发生了什么事。那个胃太大了，连比他大近两倍的鸟儿几乎都不可能有那么大的胃。

There were other odd things about him, too—for instance, the comb he wore on his middle toe-nail. What he did with it, I can´t say. He didn´t seem to do very much with his feet anyway. They were rather feeble little things, and he never used them in carrying home anything he caught. He didn´t even use them as most birds do when they stop to rest; for, instead of sitting on a twig when he was not flying, he would settle as if lying down. Sometimes he stayed on a large level branch, not cross-wise like most birds, but the long way; and when he did that, he looked like a humpy knot on the branch. When there were no branches handy, he would use a rail or a log or a wall, or even the ground; but wherever he settled himself, he looked like a blotch of light and dark, and one could gaze right at him without noticing that a bird was there. That was the way Mother Nomer did, too—clowns both of them and always ready for the wonderful game of camouflage!

他身上的奇怪事还多着呢。比如，戴在爪子中趾上的那个“梳子”。他用这个“梳子”干什么，我不知道，他好像不大用他的脚爪，他的脚爪小而无力，捉猎物回家的时候，他从来都不用脚爪；大多数鸟儿停下来休息的时候要用脚爪站立，可是他却不用，因为他落树枝上时，不是站在那里，却好像是匍匐在那里；大多数的鸟儿在树上的大横枝上栖息时，是横立在树枝上的，可是他却是纵躺着的，身体方向与树枝平行。这个时候，他看上去就像是树枝上隆起的树节；如果四周没有树，他也会选择栏杆、木头或者墙，有时候甚至是地面。总之，无论落在哪里，他看上去都像是一堆深浅不一的斑点，人们哪怕是直盯着他，也不会发现那是一只鸟儿。“错名鸟”妈妈也是这样，所以他俩都是小丑演员，总是随时准备施展他们那奇妙的伪装术！

They had remarkable voices. There seemed to be just one word to their call. I am not going to tell you what that word is. There is a reason why I am not. The reason is, that I do not know. To be sure, I have heard nighthawks say it every summer for years, but I can´t say it myself. It is a very funny word, but you will have to get one of them to speak it for you!

他们的叫声很特别，说来说去好像只有一个词。我可不会告诉你是哪个词。我不说是有原因的。那个原因就是：我不知道。当然了，很多年了，每年夏天我都听到夜鹰念叨这个词，可是我却学不来。那个词很有意思，你要想知道，就得去捉一只来让他自己说给你听。

They came by all their different kinds of queerness naturally enough, Mis and Mother Nomer did, for it seemed to run in the family to be peculiar, and all their relatives had oddities of one kind or another. Take Cousin Whip-poor-will, who wears whiskers, for instance; and Cousin Chuck-will's widow, who wears whiskers that branch. You could tell from their very names that they would do uncommon things. And as for their more distant relatives, the Hummingbirds and Chimney Swifts, it would take a story apiece as long as this to begin to tell of their strange doings. But it is a nice, likable sort of queerness they all have; so very interesting, too, that we enjoy them the better for it.

他们的这些各种各样稀奇古怪的行为举止来得都很自然,“错名鸟”先生和“错名鸟”夫人都这样,标新立异好像是流淌在家族血液里的天性,他们的近亲都有这样那样的怪癖。比如他们那长着胡须的堂兄三声夜鹰,还有胡子分叉的堂兄卡罗琳夜鹰。你光听名字就知道他们做事不同凡响。至于他们的远亲蜂鸟和烟囱雨燕,要是从头开始讲他们那古怪的所作所为,每个故事都得像这个故事一样长。不过,他们的古怪行为一点儿都不令人生厌,相反非常招人喜欢,而且非常有趣,让我们对他们更是喜爱有加。

There is one more wonderful thing yet that Mis and his mate did—and their twins with them; for before this happened, the children had grown to be as big as their parents, and a bit plumper, perhaps, though not enough to be noticed under their feathers. Toward the end of a pleasant summer, they joined a company of their kind, a sort of traveling circus, and went south for the winter. Just what performances they gave along the way, I did not hear; but with a whole flock of flying clowns on the wing, it seems likely that they had a gay time of it altogether!

“错名鸟”夫妇还做了一件了不起的事情呢，是和双胞胎一起做的。做这件事之前，两个小家伙已经和他们的父母一样高了，可能比他们稍胖一些，不过穿着羽衣看不出来。愉快的夏季即将结束的时候，他们加入了同伴的旅行马戏团，到南方去过冬了。沿途他们到底做了什么表演，我没有听见，不过，这么多飞翔小丑一起成群而行，他们玩得一定很开心！

10

The Lost Dove
One Thousand Dollars ($1000) Reward
寻鸽启事
奖金一千美元(1000 美元)

That is the prize that has been offered for a nesting pair of Passenger Pigeons. No one has claimed the money yet, and it would be a great adventure, don´t you think, to seek that nest? If you find it, you must not disturb it, you know, or take the eggs or the young, or frighten the father-or mother-bird; for the people who offered all that money did not want dead birds to stuff for a museum, but hoped that someone might tell them where there were live wild ones nesting.

这笔奖金是为寻找一对正在筑巢的候鸽设立的，至今无人认领。你一定知道去寻那个鸽巢要经历千难万险，对吧？如果你找到了，千万不要打扰他们，不要拿走鸽蛋或抓走小鸽子，也不要吓着了鸽爸爸和鸽妈妈，提供那一大笔赏金的人们可不想拿死鸟去做博物馆里的标本，他们是希望有人能告诉他们哪里有活生生的巢居的野候鸽。

You see the news had got about that the dove that is called Passenger Pigeon was lost. No one could believe this at first, because there had been so very many—more than a thousand, more than a million, more than a billion. How could more than a billion doves be lost?

你瞧，那种叫候鸽的鸽子不见了，这个消息已经不胫而走。起初，没人相信这是真的，因为以前还有很多候鸽呢，有一千只还多，一百万只以上，甚至十亿只以上。十亿多只候鸽怎么会不见了呢？

They were such big birds, too—a foot and a half long from tip of beak to tip of tail, and sometimes even longer. Why, that is longer than the tame pigeons that walk about our city streets. How could doves as large as that be lost, so that no one could find a pair, not even for one thousand dollars to pay him for the time it took to hunt?

再说了，他们可是大鸟呀，从嘴巴尖到尾巴梢有一英尺半（大约 47 厘米）长呢，有时候比这还长。咦，这么说候鸽比在城市街道上闲庭信步的驯鸽要长。这么大的鸽子怎么会不见了呢？怎么就连一对都找不到了呢？就算悬赏一千元也找不来呀？

Their colors were so pretty—head and back a soft, soft blue; neck glistening with violet, red, and gold; underneath, a wonderful purple red fading into violet shades, and then into bluish white. Who would not like to seek, for the love of seeing so beautiful a bird, even though no one paid a reward in money?

他们的色彩真是太美了：头和背是柔柔的蓝色，脖颈上是闪闪烁烁的紫色、红色和金色；腹部的颜色由深入浅逐渐过渡，开始是美丽的紫红色，逐渐变淡，变成了紫罗兰色，最后变成了青幽幽的白色。谁不愿意去寻、去找呢？即使没有金钱的奖赏，为了一饱眼福看看如此美丽的鸟儿也行呀。

Shall we go, then, to Kentucky? For it was there the man named Audubon once saw them come in flocks to roost at night. They kept coming from sunset till after midnight, and their numbers were so great that their wings, even while still a long way off, made a sound like a gale of wind; and when close to, the noise of the birds was so loud that men could not hear one another speak, even though they stood near and shouted. The place where Audubon saw these pigeons was in a forest near the Green River; and there were so many that they filled the trees over a space forty miles long and more than three miles wide. They perched so thickly that the branches of the great trees broke under their weight, and went crashing to the ground; and their roosting-place looked as if a tornado had rushed through the forest.

那么,我们就去肯塔基州吧?因为有一次,一个叫奥杜邦的人就是在那里看到他们成群结队飞来栖息过夜的。从日落时分到午夜之后，他们源源不断地涌来,数量之多连百里之外都能听到他们那狂风般的振翅声。到了近处,他们鸣声震天,人们就算是站在近处大声喊叫，也听不见对方说的话。奥杜邦是在格林河附近的森林里见到这些鸽子的。鸽群的数量之大，把森林里一片长达四十英里、宽达三英里的树木占了个满满当当。他们密密麻麻地挤在一起，大树的树枝不堪重负，被压断了，咔嚓咔嚓地落到了地上。结果，他们所到之处的森林看上去就好像是突然遭了龙卷风的侵袭一样。

Must there not be wild pigeons, yet, roosting in Kentucky—some small flock, perhaps, descended from the countless thousands seen by Audubon? No, not one of all these doves is left, they tell us, in the woods in that part of the country. The rush of their wings has been stilled and their evening uproar has been silenced. Men may now walk beside the Green River, and hear each other though they speak in whispers.

没准肯塔基州还有栖息的野鸽子呢，说不定是一小群，是奥杜邦见过的那些成千上万数不清的鸽子的后代？没有！他们说那里的森林里连一只这样的鸽子都没有了。他们匆匆挥动的翅膀已经戛然而止，他们傍晚的喧嚣鸣叫也已静默消失。现在，人类可以在格林河畔散步，即使悄声细语也能听见了。

Would you like to seek the dove in Michigan in May? For there it was, and then it was, that these wild pigeons nested, so we are told by people who saw them, by hundreds of thousands, or even millions. They built in trees of every sort, and sometimes as many as one hundred nests were made in a single tree. Almost every tree on one hundred thousand acres would have at least one nest. The lowest ones were so near the ground that a man could reach them with his hand.

你愿意五月份去密歇根州寻觅候鸽吗？因为在那里有，当时真的有，他们成千上万，甚至是数百万地聚在一起筑巢，见过他们的人都是这么说的。他们在各种各样的树上筑巢，有时候，单单一棵树上的鸽巢就多达一百个；方圆十万英里，几乎每棵树上都有至少一个鸽巢。最低的鸽巢离地面很近，人伸手就能够得着。

Suppose you should find, next May, just one pair nesting. Sire Dove, we think from what we have read, would help bring some twigs, and Dame Dove would lay them together in a criss-cross way, so that they would make a floor of sticks, sagging just a little in the middle. As soon as the floor of twigs was firm enough,

Suppose you should find just one pair.
假设你刚好发现一对候鸽。

so that an egg would not drop through, Dame Dove would put one in the shallow sagging place in the middle.

假设明年五月你只发现了一对候鸽，他们正在筑巢。那么根据我们读到的有关情况判断,鸽先生会帮着往家里拿小树枝,而鸽夫人则会把这些小树枝纵横交错地摆在一起,搭成一个中间微陷的地板。小树枝地板刚一搭好,放一个蛋也不会掉下去的时候,鸽夫人就在中间那浅浅的小凹里产了一枚蛋。

It would be a white egg, very much like those our tame pigeons lay; and, because there would be no thick soft warm rug of dried grass on the floor, you could probably see it right through the nest, if you should stand underneath and look up. But you couldn´t see it long, because, almost as soon as it was laid, Dame Dove would tuck the feather comforter she carried on her breast so cosily about that precious egg, that it would need no other padding to keep it warm.

那会是一枚白色的蛋,与驯鸽下的蛋非常相似。由于地板上没有铺又厚又软又暖和的干草毯,所以,假如你站在下面抬头看,你就能透过鸽巢看到那枚蛋。但是你看不了多久，因为那枚珍贵的蛋刚一产出来，就被鸽夫人胸前的那个羽毛被舒舒服服地裹了起来，所以，为它保暖根本不需要其他垫子。

She would stay there, the faithful mother, from about two o´clock each afternoon until nine or ten o´clock the next morning. She would not leave for one minute, to eat or get a drink of water. Then, about nine or ten o´clock each morning, Sire Dove would slip onto the nest just as she moved off, and they would make the change so quickly that the egg could not even get cool. That one very dear egg would need two birds to take care of it, one always snuggling it close while the other ate and flew about and drank.

忠实的鸽妈妈会待在那里，从午后两点左右一直待到第二天早上九十点钟，其间她不吃不喝，一刻也不离开。接下来的每天早晨大约九十点钟，她一离巢，候鸽先生就溜进去。他们交班如此之快，那枚蛋连一丝凉意都感觉不到。那枚珍贵的蛋需要两只鸟的悉心呵护，一只鸟吃东西、喝水或者飞来飞去的时候，另一只鸟则一直把它紧紧地拥在怀里。

So they would sit, turn and turn about, for fourteen days. All this while they would be very gentle with each other, saying softly, "Coo-coo," something as tame pigeons do, only in shorter notes, or calling, "Kee-kee-kee." And sometimes Sire Dove would put his beak to that of his nesting mate and feed her, very likely, as later they would feed their young. For when the two weeks´ brooding should be over, there would be a funny, homely, sprawling, soft and wobbly baby dove within the nest.

就这样，他们轮流交替，一坐就是十四天。在这十四天里，他们十分恩爱，和驯鸽一样，他们“咕咕咕”地说话，十分温柔，只不过声音短一些而已，或者他们也会“叽、叽、叽”地叫。有时候，候鸽先生还会把嘴伸到候鸽夫人的嘴跟前，给她喂食，就像以后给自己的孩子喂食那样。这很有可能哟，因为两周的孵蛋期一旦结束，巢里面就会有一个怪怪的、丑丑的、软软乎乎、滚来滚去、摇摇摆摆的候鸽宝宝了。

The father and mother of him would still have much to do, it seems; for hatching a dove out of an egg is only the easier half of the task. The wobbly baby must be brought up to become a dove of grace and beauty. That would take food.

不过，小宝宝的爸爸和妈妈要做的事情好像还多着呢。把小候鸽从蛋里孵出来仅仅完成了任务的一半，而且还是比较容易的那一半。他们必须把这个摇摇摆摆的小宝宝抚养成一只优雅美丽的鸽子。那可是需要食物的。

But you must not think to see Sire and Dame Dove come flying home with seeds or nuts or fruit or grain or earthworms or insects in their beaks. What else, then, could they bring? Well, nothing at all, indeed, in their beaks; for the food of a baby dove requires especial preparation. It has to be provided for him in the crop of his parent. So Dame Dove would come with empty beak but full crop, and the baby would be fed. Just exactly how, I have not seen written by those people who saw a million Passenger Pigeons. Perhaps they did not stop to notice.

可是，你一定不要以为你会看到候鸽先生和候鸽夫人回家的时候嘴里叼着的不是种子，就是坚果、水果、谷粒、蚯蚓或小虫儿。那么他们还能带别的什么东西呢？嗨，其实，他们嘴里什么也没有。候鸽宝宝要吃的食物很特殊，得把他们装在父母的嗉囊内才行，所以鸽夫人回家时嘴里虽然空空如也，但是嗉囊却满满当当，这下可以喂宝宝吃饭了。可是到底是怎么喂的呢？那些见过上百万只候鸽的人没有记载，也许是他们没有停下来观察吧？

However, if you will watch a tame pigeon feed its young, you can guess how a wild one would do it. A tame mother-pigeon that I am acquainted with comes to her young (she has two) and, standing in or beside the nest, opens her beak very wide. One of her babies reaches up as far as he can stretch his neck and puts his beak inside his mother´s mouth. He tucks it in at one side and crowds in his head as far as he can push it. Then the mother makes a sort of pumping motion, and pumps up soft baby food from her crop, and he swallows it. Sometimes he keeps his beak in his mother´s mouth for as long as five minutes; and if anything startles her and she pulls away, the hungry little fellow scolds and whines and whimpers in a queer voice, and reaches out with his teasing wings, and flaps them against her breast, stretching up with his beak all the while and feeling for a chance to poke his head into her mouth again. And often, do you know, his twin sister gets her beak in one side of Mother Pigeon's mouth while he is feeding at the other side, and Mother just stands there and pumps and pumps. The two comical little birds, with feet braced and necks stretched up as far as they can reach, and their heads crowded as far in as they can push them, look so funny they would make you laugh to see them. Then, the next meal Father Pigeon feeds them the same way, usually one at a time, but often both together.

不过,假如你观察一下家养驯鸽是如何给宝宝喂食的,你就能猜到野生候鸽是怎么做的了。我熟悉的一只驯鸽妈妈飞到她的宝宝跟前（她有两个宝宝），她站在巢里面或者巢旁边,把嘴巴张得大大的。一个宝宝使劲伸着脖子,把嘴伸进了妈妈的嘴里,他一边把嘴巴从妈妈嘴巴的一侧塞进去,一边还伸着小脑袋拼命地往里面挤。接着,妈妈往外一吐,软软的婴儿食品就从嗉囊里出来了,一下子就被他咽了下去。有时,他的嘴巴在妈妈的嘴巴里一待就是五分钟。假如妈妈受了惊吓,一下子缩回了嘴巴,饥饿难耐的小家伙就会又哭又怨地叨叨个没完,那声音古怪极了,同时他还伸出翅膀,拍打妈妈的前胸,缠着妈妈不放。与此同时,他一直伸着嘴巴,寻找机会再次把脑袋塞到妈妈的嘴里。你知道吗?他在妈妈嘴巴的这一侧吃食的时候，他的双胞胎妹妹常常把嘴巴塞进妈妈嘴巴的另一侧，而候鸽妈妈只是站在那里，吐呀吐，不断地往外吐食。两只滑稽可爱的小宝宝，踮着双脚、使劲伸着脖子往上够，小脑袋也使劲地往里挤，那样子实在是太有趣了，你看了一定会被逗笑的。候鸽爸爸也是这样给他们喂食的，一般情况下他是一只一只地喂，但更多的时候是两只一起喂。

Now, I think, don´t you, because that is the way tame Father and Mother Pigeon serve breakfast and dinner and supper and luncheons in between whiles to their tame twins, that wild Dame and Sire Dove would give food in very much the same way to their one wild baby? It might not be exactly the same, because tame pigeons and wild Passenger Pigeons are not the same kind of doves; but they are cousins of a sort, which means that they must have some of the same family habits.

既然驯鸽爸爸和驯鸽妈妈是这样给他们的驯鸽宝宝喂早餐、午餐、晚餐，还有中间的加餐的，所以我想，候鸽先生和候鸽夫人给他们的独生候鸽宝宝喂食的方法也差不多，你觉得呢？可能不会完全一样，因为家养驯鸽和野生候鸽不是同一类鸽子。不过，他们是近亲，所以他们一定有一些相同的家庭习惯。

If you should find a nest in Michigan in May, perhaps you can learn more about these matters, and watch to see whether, when the baby dove is all feathered out, Dame or Sire Dove pushes it out of the nest even before it can fly, though it is fat enough to be all right until it gets so hungry it learns to find food for itself. Perhaps you can watch, too, to see why Dame and Sire Dove seem to be in such a hurry to have their first baby taking care of himself. Is it because they are ready to build another nest right straight away, or would Dame Dove lay another egg in the same nest? Tame Mother Pigeon often lays two more eggs in the next nest-box even before her twins are out of their nest. Then you may be sure Father and Mother Pigeon have a busy time of it feeding their eldest twins, while they brood the two eggs in which their younger twins are growing.

假如你在五月的密歇根州找到了候鸽巢，也许这样的事情你能了解更多，而且你可以观察一下，是不是那个候鸽宝宝全身的羽毛刚长出来，甚至还不会飞的时候，候鸽先生和候鸽夫人就把他从巢里推了出去？不过，他很胖，足以支撑到饥肠辘辘时，不得不学会自己去觅食。也许你还可以观察一下，看看候鸽先生和候鸽夫人为何这样迫不及待地想让他们的第一个宝宝自己照顾自己？难道是因为他们立刻就要出发去另外筑巢吗？还是因为候鸽夫人要在这个巢里另产一个蛋？驯鸽妈妈甚至常常在她的双胞胎宝宝还没离巢的时候，就在隔壁的巢箱里再产两枚蛋。所以，毫无疑问，驯鸽爸爸和驯鸽妈妈一定忙得不亦乐乎了，他们一边要给那对大双胞胎喂食，一边还得孵化正在另外两枚蛋里生长发育的小双胞胎。

It would be very pleasant if you could watch a pair of Passenger Pigeons and find out all these things about them. *If you could!* But I said only "perhaps", because the people who know most about the matter say that Michigan has lost more than a million, or possibly more than a billion, doves. They say that, if you should walk through all the woods in Michigan, you would not hear one single Passenger Pigeon call, " Kee–kee–kee" to his mate, or hear one pair talk softly together, saying, "Coo–coo". There are sticks and twigs enough for their nests lying about; but through all the lonesome woods, so we are told, there is not one Sire Dove left to bring them to his Dame; and never, never, never will there be another nest like the millions there used to be.

假如你能仔细观察一对候鸽，并且去发现诸如此类有关他们的事情，那将是多么愉快的一件事情呀！“假如你能！”可是，我说的是“也许”，因为最了解情况的人们说，密歇根州已经失去了一百多万只候鸽，或者有可能是十亿只以上。他们说，即使你能走遍密歇根州所有的森林，你也不会听到任何一只候鸽叫伴侣时发出的“叽、叽、叽”的叫声，你也不会听到任何一对候鸽在一起柔声交谈时发出的“咕咕”声；那里到处散落着可以用来筑巢的细木棍和小树枝，可是，人们说，即使走遍整个寂静的森林，你也找不到一只能把这些细木棍和小树枝带给候鸽夫人的候鸽先生了；那曾经的数百万候鸽巢不会再有了，那样的巢永永远远也不会再有了，连一个都不会再有了。

Well, then, if we cannot find them at sunset in their roosting-place in Kentucky or in their nests in Michigan in May, shall we give up the quest for the

Through all the lonesome woods there is not one dove.
即使走遍整个寂静的森林,也找不到一只候鸽。

lost doves? Or shall we still keep hold of our courage and our hope and try elsewhere?

既然日落时，我们不能在肯塔基州他们的栖息地找到他们，既然五月里，我们不能在密歇根他们的巢窝里找到他们，那么我们是不是要放弃希望，不再寻找这些丢失的候鸽了呢？或者，我们是不是还要继续鼓足勇气，满怀希望地去别处寻找呢？

Surely, if there are any of these birds anywhere, they must eat food! Shall we seek them at some feeding–place? This might be everywhere in North America, from the Atlantic Ocean as far west as the Great Plains. That is, everywhere in all these miles where the things they liked to eat are growing. So, if you keep out of the Atlantic Ocean, and get someone to show you where the Great Plains are, you might look—*almost anywhere*. Why, many of you would not need to take a steam–train or even a trolley–car. You could walk there. Most of you could. You could walk to a place where they used to stop to feed. Those that were behind in the great flock flew over the heads of all the others, and so were in front for a while. In that way they all had a chance at a well–spread picnic ground.

Yes, you could easily walk to a place where that used to happen—most of you could.

当然了，无论哪里，如果有这种鸟，他们都必须觅食！要不我们就去觅食地找找看？北美洲到处都是觅食地，从大西洋往西一直到中央大平原。也就是说，在那绵延数千英里的地方到处是他们喜欢吃的东西。所以，假如你背对大西洋，找人指给你看大平原在哪里，你看到的就是——几乎所有的地方。嗨，你们很多人连蒸汽火车都不用坐，甚至连有轨电车都不用坐。你步行就能到那里。你们大多数人都能。你们步行就能到达他们曾经停下觅食的地方。那些落在浩浩荡荡的鸟群后面的候鸽从其他所有候鸽的头顶飞过，暂时赶到了队伍的前面，这样所有的候鸽就都有机会分享那广阔无边的野餐地了。是呀，你能轻松地步行到曾经发生过那一幕的地方——你们大多数人都能。

Do you know where acorns grow, or beechnuts, or chestnuts? Well, Passenger Pigeons used to come there to eat, for they were very fond of nuts! Do you know where elm trees grow wild along some riverway, or where pine trees live? Oh! that is where these birds used sometimes to get their breakfasts, when the trees had scattered their seeds. Do you know a tree that has a seed about the right size and shape for a knife at a doll´s tea-party? Yes, that´s the maple; and many and many a party the Passenger Pigeons used to have wherever they could find these cunning seed-knives. Only they didn´t use them to cut things with. They ate them up as fast as ever they could.

你知道哪里有橡子、山毛榉坚果或栗子吗？嗯，候鸽过去常常飞到那里去觅食，因为他们非常喜欢吃坚果！你知道哪里的河边有疯长的榆树，或者哪里有松树吗？噢！那些树的种子四处散落时，这些鸟儿有时候也去那里吃早餐。你知道一种树吗？这种树有一个种子，种子的大小和形状跟玩具娃娃茶话会上的小刀差不多。对了，那就是枫树。过去，只要发现这些精巧可爱的“种子小刀”，他们都会举办一次宴会，已经举办过无数次了。只不过他们不用这些“种子小刀”切东西，而是赶着把他们吃光，能吃多快就吃多快。

Have you ever picked wild berries? Why, more than likely Passenger Pigeons have picked other berries there or thereabouts before your day!

你摘过野浆果吗？嗨，在你之前，候鸽很可能早就已经在那里或者周围摘过其他浆果了!

Do you know a place where the wild rice grows? Ah, so did the Passenger Pigeons, once upon a time!

你知道哪个地方有野生稻米吗?啊,很久以前,候鸽也知道!

But if you know none of these places, even then you can stand near where the flocks used to fly when they were on their journeys. All of you who live between the Atlantic Ocean and the Great Plains can go to the door or a window of the house you live in and point to the sky and think: "Once so many Passenger Pigeons flew by that the sound of their wings was like the sound of thunder, and they went through the air faster than a train on a track, and the numbers in their flocks were so many that they hid the sun like great thick clouds."

就算这些地方你连一个都不知道,你还可以靠近候鸽群旅途中曾经飞过的地方。你们都住在大西洋和大平原之间,所以你们都可以走到自家门前或窗前,面向天空，默默地想："曾经有那么多候鸽从这里飞过，他们挥动翅膀，响声震天，就像是那隆隆的雷鸣；他们从天空飞过，速度比轨道上的火车还要快；他们结伴而行、数量庞大，像大块厚厚的乌云遮天蔽日。"

When you do that, some of you will doubtless see birds flying over; but we fear that not even one of you will see even one Passenger Pigeon in its flight.

这样做的时候，你们中的有些人肯定会看到飞过天空的鸟儿，可是，我恐怕你们当中不会有哪怕一个人看到哪怕是一只展翅而飞的候鸽。

What happened to the countless millions is recorded in so many books that it need not be written again in this one. This story will tell you just one more thing about these strange and wonderful birds, and that is that no *child* who reads this story is in any way to blame because the dove is lost. What boy or girl is not glad to think, when some wrong has been done or some mistake has been made, "It´s not *my* fault"?

数不清的成百上千万的候鸽到底遭遇了怎样的不

幸？已经有太多的书对此进行了记录，所以在此就无须再说了。关于这些奇异而美丽的鸟儿，这个故事只想再告诉你一件事情，这件事情就是：无论如何，我们不能因为失去了候鸽而责怪阅读这个故事的任何一个小朋友。只要是孩子，面对自己的过失或错误，哪个不乐意说："这不是我的错"呢？

Even though this bird is gone forever and forever and forever, there are many other kinds living among us. If old Mother Earth has been robbed of some of her children, she still has many more—many wonderful and beautiful living things. And that she may keep them safe, she needs your help; for boys and girls are her children, too, and the power lies in your strong hands and your courageous hearts and your wise brains to help save some of the most wonderful and fairest of other living things. And what one among you all, I wonder, will not be glad to think that *you* help keep the world beautiful, when you leave the water-lilies floating on the pond; that it is the same as if *you* sow the seeds in wild gardens, when you leave the cardinal flowers glowing on the banks and the fringed gentians lending their blue to the marshes. For the life of the

Once, so many flew by, that the sound of their wings was like the sound of thunder.
它们挥动翅膀，响声震天，就像是那隆隆的雷鸣。

world, whether it flies through the air or grows in the ground, is greatly in your care; and though you may never win a prize of money for finding the dove that other people lost, there is a reward of joy ready for anyone who can look at our good old Mother Earth and say, "It will not be *my* fault if, as the years go by, you lose your birds and flowers."

尽管候鸽已经永远、永远地消失了，再也不会来了，可还有许多各种各样其他的鸟儿和我们生活在一起。虽然大自然老妈妈已经痛失了一些孩子，可是，她还有其他很多孩子，他们是那么神奇而又美丽！为了让他们安安全全、不受伤害，她需要你的帮助，无论你是男孩子还是女孩子，都是大自然妈妈的孩子，拯救那些最奇妙、最美丽的生命，需要你的力量，这力量就握在你那双有力的手上，就埋在你那颗勇敢的心里，就藏在你那个聪明的大脑中。我想知道，那将是你们中间的谁呢？看着睡莲在池塘上轻轻荡漾而不去打扰，心里快乐地想着原来“我”也可以让这个世界如此美丽！同样，当“你”任由红花半边莲在岸边鲜艳欲滴，任由蓝莹莹的龙胆花在沼泽地里瞎涂乱抹的时候，你也是在为这个美丽的世界贡献自己的一份力量，因为是“你”把种子播撒在了茫茫的野生花园里。在这个世界里，所有的生命，无论是天空中飞的，还是地上长的，都仰仗着你的呵护。也许你永远也不会找到别人遗失的那只候鸽，得到金钱的奖赏，可是，无论是谁，只要能看着我们慈爱的大地老妈妈，对她说："假如岁月流逝，您失去了鸟儿和花儿，那不会是我的错"，那么这个人就会得到奖赏，那奖赏就是他发自内心的喜悦！

And it would be, don´t you think, one of the greatest of adventures to seek and find and help keep safe such of these as are in danger, that they may not, like the dove, be lost?

寻觅那些濒临灭绝的生命，找到他们，为他们的安全助一臂之力，让我们不要像失去候鸽那样再失去他们，那将是人生最最难忘的经历，你说呢？

11

Little Solomon Otus
小猫头鹰所罗门

Oh, the wise, wise look of him, with his big round eyes and his very Roman nose! He had sat in a golden silence throughout that dazzling day; but when the kindly moon sent forth a gentler gleam, he spoke, and the speech of little Solomon Otus was as silver. A quivering, quavering whistle thrilled through the night, and all who heard the beginning listened to the end of his song.

哇,真是一张聪明绝顶的脸:大大的圆眼睛,还有那高高的直鼻梁! 他已经在那里坐了一整天了,耀眼的日光下,他一言不发,难开金口。可是,当善解人意的月亮洒下柔柔的月色时，他开口说话了。小猫头鹰所罗门开口就是银，那是颤了又颤、抖了又抖的口哨声，那声音划破夜空的时候，无论是谁只要听到了，都会从头听到尾。

It was a night and a place for music. The mellow light lay softly over the orchard tree, on an old branch of which little Solomon sat mooning himself before his door. He could see, not far away, the giant chestnut trees that shaded the banks of a little ravine; and hear the murmuring sound of Shanty Creek, where Nata grew up, and where her grandchildren now played hide-and-seek. Near at hand stood a noble oak, with a big dead branch at the top that was famous the country round as a look-out post for hawks and crows; and maybe an eagle now and then had used it, in years gone by.

这是个充满音乐的夜晚，这是个充满音乐的地方。柔和的月光透过果园里的果树，轻轻地洒在一棵老树枝上。就在这棵老树枝上，小所罗门正坐在自家的门前神思遐想呢。瞧，不远处的那棵栗子树遮天蔽月，在一条小沟边投下了层层暗影；听，“木屋溪”在潺潺低语，蜻蜓娜塔就是在那里长大的呀，现在她的孙儿、孙女们正在那里捉迷藏呢。距他咫尺之遥的地方,矗立着一棵高大挺拔的橡树,树顶有一棵远近闻名的大枯枝,那是鹰和乌鸦们的瞭望哨;在那遥远的岁月里,也许有鹰偶尔来这里瞭望过吧?

But hawk and crow were asleep, and toads were trilling a lullaby from the pond, while far, far off in the heart of the woods, a whip-poor-will called once, twice, and again.

可是,现在鹰和乌鸦都睡着了;池塘里蛤蟆咯呱咯呱地唱着催眠曲;远远的森林深处,三声夜鹰叫了第一声,接着又叫了第二声,然后又叫了第三声。

Solomon loved the dusk. His life was fullest then and his sight was keenest. His eyes were wide open, and he could see clearly the shadow of the leaves when the wind moved them lightly from time to time. He was at ease in the great night-world and master of many a secret that sleepy-eyed day-folk never guess. As he shook out his loose, soft coat and breathed the cool air, he felt the pleasant tang of a hunger that has with it no fear of famine.

所罗门太喜欢黄昏的幽暗了！只有那个时候，他的生活最充实，他的视力也最敏锐。他的眼睛总是瞪得大大的，连偶尔风吹树叶，轻轻投在地上的影子都看得一清二楚。在茫茫的夜的世界里，他轻松自在、应付自如，而且他掌握着许多秘密，那些秘密是一到晚上就睡眼蒙眬的"白猫子"们永远也猜不出来的。他抖抖身上宽松、柔软的大衣，吸一口凉凉的风，这时，他感觉到了一丝饥饿，这让他欣喜兴奋，所罗门的饥饿感是不会带给他饥肠辘辘的恐惧的。

Once more he sent his challenge through the moonlight with quivering, quavering voice, and some who heard it loved the darkness better for this spirit of the night, and some shivered as if with dread. For Solomon had sounded his hunting call, and, as with the baying of hounds or the tune of a hunter´s horn, one ear might find music in the note and another hear only a wail.

他那颤了又颤、抖了又抖的声音又一次划破了朦胧的月夜，那是他发出的挑战，有些人听到了，愈加喜爱这茫茫的黑夜了，他们喜欢这夜的精灵；有些人却不由自主地打起了寒战，好像很害怕的样子。所罗门已经吹响了狩猎的号角，正如猎犬的狂吠，或者猎人的号角，有的人听到的是乐符，有些人听到的却仅仅是哀号。

Then, silent as a shadow, he left his branch. Solomon, a little lone hunter in the dark, was off on the chase. Whither he went or what he caught, there was no sound to tell, until, suddenly, one quick squeak way over beside the corn-crib might have notified a farmer that another mouse was gone. But the owner of the corn-crib was asleep, and dreaming, more than likely, that the cat, which was at that moment disturbing a pair of meadow bobolinks, was somehow wholly to be thanked for the scarcity of mice about the place.

接着，悄无声息，就像影子一样，他离开了树枝。所罗门，这个黑暗中的独行小猎手，他出发去捕猎了。他去了哪里？捉住了什么？一切都静悄悄的，没人能听见。可是，突然，远处谷仓旁传来"吱"的一声尖叫，声音短促，稍纵即逝，于是农夫可能就知道了：又有一只老鼠毙命了。可是，谷仓的主人却在熟睡，而且，非常有可能正在做梦，在梦中，他还不知在怎么感谢他家的猫呢，因为他以为谷仓四周老鼠这么少，全托那只猫的福了；

可实际上，此时此刻，那只猫正在草地上骚扰一对食米鸟呢。

Solomon was not wasteful about his food. He swallowed his evening breakfast whole. That is, he swallowed all but the tail, which was fairly long and stuck out of his mouth for some time, giving him rather a queer two-tailed look, one at each end! But there was no one about to laugh at him, and it was, in some respects, an excellent way to make a meal. For one thing, it saved him all trouble of cutting up his food; and then, too, there was no danger of his overeating, for he could tell that he had had enough as long as there wasn´t room for the tail. And after the good nutritious parts of his breakfast were digested, he had a comfortable way of spitting out the skin and bones all wadded together in a tidy pellet. An owl is not the only kind of bird, by any means, that has a habit of spitting out hard stuff that is swallowed with the food. A crow tucks away many a discarded cud of that sort; and even the thrush, half an hour or so after a dainty fare of wild cherries, taken whole, drops from his bill to the ground the pits that have been squeezed out of the fruit by the digestive mill inside of him.

Oh, the wise, wise look of him.
哇，真是一张聪明绝顶的脸。

所罗门吃东西的时候从不浪费。傍晚的“早餐”他整个儿一口就吞了下去。也就是说，除了尾巴以外，他都一口吞下去了，因为尾巴有点太长了，有一会儿的工夫，一直露在他的嘴巴外面没咽下去，结果他看上去像长了两个尾巴的怪物，屁股那里有一条尾巴，嘴巴那里竟然还有一条尾巴！不过，周围没人，所以不会有人笑话他。从有些方面来看，这种吃法有很多好处。一方面，省去了切食物的所有麻烦，另一方面，没有过度饮食的危险，因为只要肚里装不下那条尾巴，他就知道自己吃饱了。等到“早餐”里那有益健康、营养丰富的部分被消化掉之后，他再轻松自如地把皮和骨头吐出来，这时，这些皮和骨头早就变成了一个圆滚滚的小球丸。无论如何，除了猫头鹰，还有其他很多鸟也有把随食物吞下去的硬东西再吐出来的习惯。乌鸦就能反复咀嚼那种吐出来的反刍食物；甚至连画眉鸟都有这个习惯。画眉吃野樱桃的时候，总是整个儿吞到肚子里，可是他体内起消化作用的研磨机会把樱桃核挤出来，所以，每次大吃一顿野樱桃后大约半个小时左右，樱桃核就会从画眉的嘴里吐出来落到地上。

After his breakfast, which he ate alone in the

evening starlight and moonlight, Solomon passed an enjoyable night; for that world, which to most of us is lost in darkness and in sleep, is full of lively interest to an owl. Who, indeed, would not be glad to visit his starlit kingdom, with eyesight keen enough to see the folded leaves of clover like little hands in prayer—a kingdom with byways sweet with the scent and mellow with the beauty of waking primrose? Who would not welcome, for one wonderful night, the gift of ears that could hear the sounds which to little Solomon were known and understood, but many of which are lost in deafness to our dull ears?

借着傍晚的星光和月色，所罗门独自一人吃完了"早餐"，接下来的夜晚愉快而又有趣。那个对我们大多数人来说意味着无边的黑暗和漫长的沉睡的世界，对猫头鹰来说，却是一个充满活力、妙趣横生的世界。的确，倘若我们视力敏锐，看得见那像小手一样合在一起做祈祷状的三叶草，那么，谁不愿意在他那星光闪烁的王国里流连忘返呢？那里，路边小径芳香四溢、美不胜收，处处是彻夜不眠的报春花呀。倘若我们的耳朵能够收到一份礼物，给我们一个美妙的夜晚，让我们听听小所罗门熟悉并且心领神会的那些声音，那么谁不愿意欣然接受呢？因为我们的耳朵已经迟钝，已经听不见其中很多的声音了。

Of course, it may be that Solomon never noticed that clovers fold their leaves by night, or that primroses are open and fragrant after dusk. For he was an owl, and not a person, and his thoughts were not the thoughts of man. But for all that they were wise thoughts—wise as the look of his big round eyes; and many things he knew which are unguessed secrets to dozy day–folk.

当然啦，所罗门有可能从未留意过三叶草的叶子一到晚上竟然会合起来，他也有可能从未留意过报春花黄昏后还会绽开笑脸、吐露芬芳。所罗门只是一只猫头鹰，他可不是人，他的所思所想也不是人类的那种所思所想。可是，尽管如此，他的所思所想都绝顶聪明，和他那大大的圆眼睛里透出的聪明一个样。他知道的很多事情，对那些一到晚上就昏昏欲睡的“白猫子”来说，是猜也猜不出来的谜。

He was a successful hunter, and he had a certain sort of knowledge about the habits of the creatures he sought. He seldom learned where the day birds slept, for he did not find motionless things. But he knew well enough that mice visited the corn –crib, and where their favorite runways came out into the open. He knew where the cutworms crept out of the ground and feasted o´nights in the farmer´s garden. He knew where the big brown beetles hummed and buzzed while they munched greedily of shade-tree leaves. And he knew where little fishes swam near the surface of the water.

他是一个成功的猎手,对猎物的习性自有其独到的了解。他很少去了解日间活动的鸟儿睡在哪里,因为他不捉静止不动的东西。不过他很清楚谷仓是老鼠经常光顾的地方, 他也很清楚它们通往野外的通道在哪里。他知道切根虫会从哪里爬出地面,到农夫家的花园里整夜大嚼特嚼;他知道哪里有棕色的大甲壳虫一边嗡嗡嘤嘤地叫着,一边贪婪地啃噬绿荫葱茏的树叶;他还知道哪里的水面会有各种各样游来游去的小鱼儿。

So he hunted on silent wings the bright night long; and though he did not starve himself, as we can guess from what we know about his breakfast of rare mouse-steak, still, the tenderest and softest delicacies he took home to five fine youngsters, who welcomed their father with open mouths and eager appetite. Though he made his trips as quickly as he could, he never came too soon to suit them-the hungry little rascals.

所以，他乘着无声的翅膀，整夜欢快地忙着捕食。尽管他不会饿着自己，这点从他“早餐”吃生鼠排的情形就看得出来，不过，最嫩、最软的美食他还是要带回家给自己的五个小宝宝吃，因为他们总是张着嘴巴、急不可待地等着爸爸的归来。虽然每次他都竭尽全力尽快返回，可是再快也赶不上他们饿的快——唉，这些总也吃不够的小调皮呀！

They were cunning and dear and lovable. Even a person could see that, to look at them. It is not surprising that their own father was fond enough of them to give them the greater part of the game he caught. He had, indeed, been interested in them before he ever saw them—while they were still within the roundish white eggshells, and did not need to be fed because there was food enough in the egg to last them all the days until they hatched.

他们好小巧，好可爱，好招人喜欢呀！这点连看到他们的人类都看得出来，所以他们的爸爸如此的喜欢他们，把最好的猎物留给他们吃就不足为怪了。其实，连他们的面都还没见过的时候，爸爸对他们就已经是关注有加了，那个时候，他们还在圆圆的白色蛋壳里没出来

Solomon knew the runways of the mice.
所罗门很清楚老鼠的通道在哪里。

呢，不需爸爸给他们喂食，因为蛋壳的食物足够他们享用到出壳的那一天。

Yes, many a time he had kept those eggs warm while Mrs. Otus was away for a change; and many a time, too, he stayed and kept her company when she was there to care for them herself. Now, it doesn´t really need two owls at the same time to keep a few eggs warm. Of course not! So why should little Solomon have sat sociably cuddled down beside her? Perhaps because he was fond of her and liked her companionship. It would have been sad, indeed, if he had not been happy in his home, for he was an affectionate little fellow and had had some difficulty in winning his mate. There had been, early in their acquaintance, what seemed to Solomon a long time during which she would not even speak to him. Why, it is said he had to bow to her as many as twenty or thirty times before she seemed even to notice that he was about. But those days were over for good and all, and Mrs. Otus was a true comrade for Solomon as well as a faithful little mother. Together they made a happy home, and were quite charming in it.

是呀，好多次猫头鹰太太不在的时候，他就为这些蛋保暖；还有好多次，太太自己待在窝里照顾蛋宝宝时，他就留下来陪她。嗨，真的没必要两只猫头鹰同时给蛋宝宝保暖。当然没必要啦！可为什么小所罗门要依偎在她身旁陪她呢？也许是因为他爱她，喜欢和她在一起吧。说实在的，要是小所罗门没能得到家庭幸福的话那可够令人伤感的，因为他是个感情丰富的小家伙，而且，为了赢得意中人的芳心他还颇费了一番周折呢。他俩刚刚认识的时候，她甚至连一句话都不跟他说，对所罗门来说那可是段漫长的日子呀。唉，据说他得不断地冲着她鞠躬，要鞠二三十次，她似乎才会注意到他的存在。好在那段日子结束了，永远结束了，现在，猫头鹰太太不仅是所罗门的亲密伴侣，还是一个尽职尽责的小妈妈。他们一起营造了一个幸福的家，相亲相爱，十分美满。

They could be brave, too, when courage was needed, as they gave proof the day that a boy wished he hadn´t climbed up and stuck his hand in at their door–hole, to find out what was there. While Mrs. Otus spread her feathers protectingly over her eggs, Solomon lay on his back, and, reaching up with beak and clutching claws, fought for the safety of his family.

In the heat of the battle he hissed, whereupon the boy retreated, badly beaten, but proudly boasting of an adventure with some sort of animal that felt like a wildcat and sounded like a snake.

必要时，他们会很勇敢，这一点已经被那个追悔莫及的小男孩证实。那天，小男孩爬上来，把手伸进了他们的家门，想看看里面是什么。猫头鹰太太立刻张开翅膀护住蛋宝宝，所罗门则腿脚冲上躺下，伸出尖嘴和利爪，为家人的安全而战。一边激烈地战斗，一边还不断地发出“嘶嘶嘶”的叫声，小男孩不得不撤退了。虽然小男孩被狠狠地揍了一顿，可是他却美滋滋的，对这次历险津津乐道，吹嘘说自己和一种摸上去像是野猫、听上去像是蛇的动物交过战。

Besides, courage when needed, health, affection, good-nature, and plenty of food were enough to keep a family of owls contented. To be sure, some folk might not have been so well satisfied with the way the household was run. A crow, I feel quite sure, would not have considered the place fit to live in. Mrs. Otus was not, indeed, a tidy housekeeper. The floor was dirty—very dirty—and was never slicked up from one week´s end to another. But then, Solomon didn´t mind. He was used to it. Mrs. Otus was just like his own mother in that respect; and it might have worried him a great deal to have to keep things spick and span after the way he had been brought up. Why, the beautiful white eggshell he hatched out of was dirty when he pipped it, and never in all his growing-up days did he see his mother or father really clean house. So it is no wonder he was rather shiftless and easy-going. Neither of them had shown what might be called by some much ambition when they went house-hunting early that spring; for although the place they chose had been put into fairly good repair by rather an able carpenter, —a woodpecker,—still, it had been lived in before, and might have been improved by having some of the rubbish picked up and thrown out. But do you think Solomon spent any of his precious evenings that way? No, nor Mrs. Otus either. They moved in just as it was, in the most happy-go-lucky sort of way.

必要时的勇气、健康、友爱、好脾气以及足够的食物，这些东西加起来足以让猫头鹰一家心满意足。当然了，有些人对他们过的日子可能不是很满意。我敢肯定，乌鸦就不喜欢他们住的地方，他肯定会认为那里不宜居住。的确，猫头鹰太太不是一个整洁的女主人。他家的地板很脏，非常脏，从来不收拾，周周如此。可是，所罗门并不介意，他已经习惯了。在这方面，他的太太和自己的妈妈一样，所以，假如真要违背小时候的生活习惯，把一切弄得整整齐齐、一尘不染的话，那会让小所罗门非常不自在的。再说了，他破壳而出的时候，那个漂亮的白蛋壳也是脏乎乎的，而且，从小到大，他从来没有见过妈妈或者爸爸真正地打扫过屋子。所以，怪不得小所罗门会这样得过且过，随遇而安呢。那年初春，去为筑巢选址的时候，他俩谁也没有表现出有人所说的那种宏图大志。因为，尽管他们选的地方已经被能干的木匠啄木鸟先生修缮一新，但毕竟已经住过了，里面还有些垃圾需要捡起来扔掉。可是，你以为所罗门会在那宝贵的夜晚收拾垃圾吗？不会的，猫头鹰太太也不会。他们直接就搬进去了，一副逍遥自在、无忧无虑的样子。

Well, whatever a crow or other particular person might think of that nest, we should agree that a father and mother owl must be left to manage affairs for their young as Nature has taught them; and if those five adorable babies of Solomon didn´t prove that the way they were brought up was an entire success from an owlish point of view, I don´t know what could.

嗨，无论乌鸦或者其他什么人对那个巢有什么意见，我们都得承认，为孩子操心的事情还是得由猫头鹰爸爸和猫头鹰妈妈自己按照大自然妈妈的教导去做。假如连所罗门那五个可爱的宝宝都无法证明猫头鹰夫妇的养育方式是成功的，那我就不知道还有什么能证明了，当然，那种成功得用猫头鹰的观点来评价。

Take them altogether, perhaps you could not find a much more interesting family than the little Otuses. As to size and shape, they were as much alike as five peas in a pod; but for all that, they looked so different that it hardly seemed possible that they could be own brothers and sisters. For one of the sons of Solomon and two of his daughters had gray complexions, while the other son and daughter were reddish brown. Now Solomon and Mrs. Otus were both gray, except, of course, what white feathers and black streaks were mixed up in their mottlings and dapples; so it seems strange enough to see two of their children distinctly reddish. But, then, one never can tell just what color an owl of this sort will be, anyway. Solomon himself, though gray, was the son of a reddish father and a gray mother, and he had one gray brother and two reddish

Those five adorable babies of Solomon.
所罗门那五个可爱的宝宝。

sisters: while Mrs. Otus, who had but one brother and one sister, was the only gray member of her family. Young or old, summer or winter, Solomon and Mrs. Otus were gray, though, young or old, summer or winter, their fathers had both been of a reddish complexion.

五个小宝宝在一起的时候，简直是太有趣了，也许你再也找不到比小猫头鹰更有趣的家庭组合了。从大小和形状上来看，他们非常相似，就如同豆荚里的五粒豌豆，大小形状都一样。可是，尽管如此，看上去他们却一点儿也不像，根本就不像是亲兄弟姐妹。因为，其中一个宝宝儿子和两个宝宝女儿灰不拉叽的，而另外一个宝宝儿子和宝宝女儿却是红彤彤的。所罗门和他的太太全身灰色，当然，还有白羽毛和黑条纹混杂而成的斑斑点点。所以，看到他们的孩子中竟然有两个穿着与众不同的红衣服，似乎就太奇怪了。可是，话说回来，这种猫头鹰身上的颜色谁都无法预测。别看所罗门自己灰不拉叽的，可他父母却不都这样。他妈妈身上的羽毛是灰的，可是，他爸爸却满身红彤彤的；而且，虽然所罗门的哥哥满身灰色，两个姐姐却满身彤红；而他的太太家兄妹三个，只有他太太身上的羽毛是灰的。总之，无论年轻还是年老，也无论是酷暑还是严冬，所罗门和猫头鹰太太都是一身灰；而无论年轻还是年老，也无论是酷暑还是严冬，他们俩的爸爸却都是一身红。

Now this sort of variation in color you can readily see is altogether a different matter from the way Father Goldfinch changes his feathers every October for a winter coat that looks much the same as that of Mother Goldfinch and his young daughters; and then changes every spring to a beautiful yellow suit, with black–and–white trimmings and a black cap, for the summer. It is different, too, from the color –styles of Bob the Vagabond, who merely wears off the dull tips of his winter feathers, and appears richly garbed in black and white, set off with a lovely bit of yellow, for his gay summer in the north. Again, it is something quite different from the color–fashions of Larie, who was not clothed in a beautiful white garment and soft gray mantle, like his father´s and mother´s, until he was quite grown up.

显然，这种色差方式与金丝雀完全不同。每年十月，金丝雀爸爸都会换上冬装，可是，他的冬装看上去与他的太太和女儿们身上的冬装差不多一样；每年春天，为了度过盛夏，他又会换上一身漂亮的黄色套服，上面还装饰着黑色和白色的边，另外，还要戴上一顶黑色的帽子。“流浪汉”食米鸟鲍勃的换色风格也不同:为了北方那绚丽多彩的夏季,鲍勃只是把冬衣羽毛尖上的暗灰色慢慢地抖掉,露出身上白色和黑色的羽毛,上面还带着一丝淡淡的黄,就好像是穿着光鲜亮丽的盛装一样;还有,这与潜鸟拉瑞的换色时尚也大相径庭,拉瑞长大后,才能像爸爸妈妈那样,穿上了漂亮的白色外套,披上了柔柔的灰色斗篷。

No, the complexion of Solomon and his sons and daughters was a different matter altogether, because it had nothing whatever to do with season of the year, or age, or sex. But for all that it was not different from the sort of color-variations that Mother Nature gives to many of her children; and you may meet now and again examples of the same sort among flowers, and insects, and other creatures, too.

的确,所罗门和他的儿女们身上的颜色问题很特殊,因为他和季节、年龄,或者性别毫无关系。可是,尽管如此,他们的这种色差方式与大自然母亲赋予其众多子女的色差方式毫无二致,你会时不时地发现这样的鲜花、这样的昆虫,还有这样的其他生物。

But, reddish or gray, it made no difference to

Solomon and Mrs. Otus. They had no favorites among their children, but treated them all alike, bringing them food in abundance: not only enough to keep them happy the night long, but laying up a supply in the pantry, so that the youngsters might have luncheons during the day.

无论是红还是灰，对所罗门和他的太太来说没什么两样。他们不偏爱，对孩子们都是一视同仁，他们为孩子们带回大量食物，不仅够他们晚上吃，让他们整个晚上都快快乐乐的，还堆起一堆，储存起来，好在白天给孩子们加餐。

Although Solomon had night eyes, he was not blind by day. He passed the brightest hours quietly for the most part, dozing with both his outer eyelids closed, or sometimes sitting with those open and only the thin inner lid drawn sidewise across his eye. It seems strange to think of his having three eyelids; but, then, perhaps we came pretty near having a third one ourselves; for there is a little fold tucked down at the inner corner, which might have been a third lid that could move across the eye sidewise, if it had grown bigger. And sometimes, of a dazzling day in winter, when the sun is shining on the glittering snow, such a thin lid as Solomon had might be very comfortable, even for our day eyes, and save us the trouble of wearing colored glasses.

虽然所罗门的眼睛是夜眼，不过白天他照样也能看清。阳光最强烈的时候，大部分时间他是安安静静地度过的，有时合着两个外眼皮打盹儿，有时睁着外眼皮坐在那里，只有薄薄的内眼皮横拉过来遮着眼睛。想想他有三个眼皮，似乎有点奇怪耶！可是，我们人类自己其实差点也会有三个眼皮呢。我们的内眼角那里有一个小褶层，如果当初这个小褶层长大了，那可能就是我们的第三个眼睑了，也能在眼睛上横着来回移动。冬天，太阳照在晶莹的雪地上明晃晃的，所以，有时候，如果有所罗门这样的一个薄眼皮，那一定会非常舒服，甚至对我们只能白天看东西的眼睛也是如此，那样就省得我们再戴墨镜了。

Lively as Solomon was by night, all he asked during the day was peace and quiet. He had it, usually. It was seldom that even any of the wild folk knew where his nest was; and when he spent the day outside, in some shady place, he didn´t show much. His big feath-

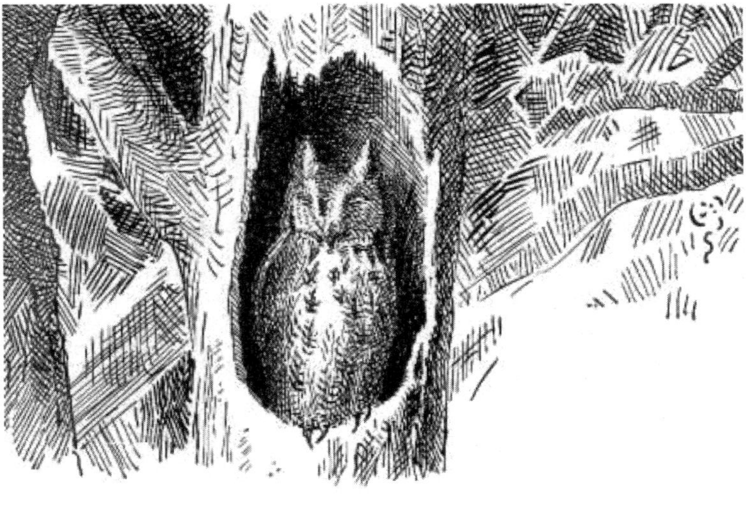

He passed the brightest hours dozing.
阳光最强烈的时候，他在打盹。

er–horns at such times helped make him look like a ragged stub of a branch, or something else he wasn´t. It is possible for a person to go very close to an owl without seeing him; and fortunately for Solomon, birds did not find him every day. For when they did, they mobbed him.

晚上，所罗门充满活力，可是一到白天，他只求安安静静地待着。通常，他都能如愿，因为就连那些整天待在野外的家伙也很少知道他的巢在哪里，而且，白天待在外面的浓荫里时，他也不怎么显眼。那时，他那像犄角一样的羽毛就会竖起来，把他伪装成一根粗巴巴的断树枝或者别的什么东西，就算你走到跟前也有可能看不见他。幸运的是，鸟儿们可不是每天都能发现所罗门的，因为，一旦发现了，他们就会群起而攻之。

One day, rather late in the summer, Cock Robin found him and sent forth the alarm. To be sure, Solomon was doing no harm—just dozing, he was, on a branch. But Cock Robin scolded and sputtered and called him mean names; and the louder he talked, the more excited all the other birds in the neighborhood became. Before long there were twenty angry kingbirds and sparrows and other feather–folk, all threatening to do something terrible to Solomon.

一天，那时已经夏末，知更鸟考克·罗宾发现了他，立刻发出了警报。说实话，当时所罗门并未伤害谁，他只是在一根树枝上打盹。可是考克·罗宾却冲着他又骂又嚷，唾沫星子乱溅，骂得别提有多难听了。他越骂声音越大，周围其他的鸟儿越听越群情激愤，不一会儿就聚集了二十多只愤怒的鸟儿，有必胜鸟，有麻雀，还有其他羽友，他们都威胁着要给所罗门点儿颜色看看。

Now, Solomon had been having a good comfortable nap, with his feathers all hanging loose, when Cock Robin chanced to alight on the branch near him. He pulled himself up very thin and as tall as possible, with his feathers drawn tight against his body. When the bird-mob got too near him, he looked at them with his big round eyes, and said, "Oh! " in a sweet high voice. But his soft tone did not turn away their wrath. They came at him harder than ever. Then Solomon showed his temper, for he was no coward. He puffed his feathers out till he looked big and round, and he snapped his beak till the click of it could be heard by his tormentors. And he hissed.

所罗门刚才还耷拉着全身的羽毛，舒舒服服地睡得香呢，可是，考克·罗宾却突然飞了过来，不小心落在了他旁边的树枝上。此刻，他羽毛紧贴身体，全身紧绷，使劲地竖起了身子。当这些聚集的鸟儿离得太近时，他就瞪起又大又圆的眼睛看着他们，“噢”地叫一声，声音很大，却还温和。可是，这温和的叫声并未驱散鸟儿们的愤怒，他们扑得更猛了。这下，所罗门火了，他可不是个懦夫。只见他乍起全身的羽毛，身体鼓得又圆又大，嘴巴不停地咬着，噼啪作响，让所有这些使他不得安宁的鸟儿们都听了个真真切切。同时，他还发出嘶嘶的叫声。

But twenty enemies were too many, and there was only one thing to be done. Solomon did it. First thing those birds knew, they were scolding at nothing at all; and way off in the darkest spot he could find in the woods, a little owl settled himself quite alone and listened while the din of a distant mob grew fainter and fainter and fainter, as one by one those twenty birds discovered that there was no one left on the branch to scold at.

可是，有二十个敌人呢，终究是寡不敌众呀。三十六计，走为上计。所罗门就如此照办了。接下来这些鸟儿发现，他们骂了半天白骂了。远处的树林里，光线最暗的一个地方，有一只小猫头鹰正休息呢，就他自己。远处那讨厌的二十只鸟儿叽叽喳喳的叫骂声越来越小，越来越小了，因为他们一个接一个地发现，树枝上根本就没有可骂的对象了。

If Solomon knew why the day birds bothered him so, he never told. He could usually keep out of their way in the shady woods in the summer; but in the winter, when the leaves were off all but the evergreen trees, he had fewer places to hide in. Of course, there were not then so many birds to worry him, for most of them went south for the snowy season.

为什么这些习惯于白天出来活动的鸟儿要如此这般地搅扰所罗门呢？就算他知道，他从来也没说过。夏天，他通常待在树林里的浓荫里，离他们远远的；可是到了冬季，除了常青树以外，所有其他树的树叶都掉光了，他也就没多少地方可躲啦。当然了，那时也不会有太多的鸟儿来烦他，因为他们大都飞到南方过冬去了。

But Jay stayed through the coldest days and enjoyed every chance he had of pestering Solomon. I don´t know that this was because he really disliked the little owl. Jay was as full of mischief as a crow, and if the world got to seeming a bit dull, instead of moping and feeling sorry and waiting for something to happen, Jay looked about for some way of amusing himself. He was something of a bully,—a great deal of a bully, in fact,—this dashing rascal in a gay blue coat; and the more he could swagger, the better he liked it.

可是讨厌鬼松鸦一直没走,即使在最冷的那些日子里,他都待在这里,逮住机会就骚扰小所罗门。我可不知道他之所以这样做是因为他真的不喜欢小猫头鹰呀。在乌鸦家族中,讨厌鬼松鸦满肚子都是坏水。假如日子平平淡淡,有点无聊,讨厌鬼松鸦不会做做鬼脸、伤心一阵,然后期待发生点什么就完事,他会这儿瞧瞧,那儿望望,想办法找点事让自己高兴。他有点像个小霸王,实际上,他霸道极了,别看他穿着华丽的蓝外套,可是骨子里却是个喜欢寻衅滋事的小恶棍,越是耀武扬威,自我感觉就越好。

He seemed, too, to have very much the same feeling that we mean by joy, in fun and frolic. There was, perhaps, in the sight of a bird asleep and listless in broad daylight, something amusing. He was in the habit of seeing the feather–folk scatter at his approach. If he understood why, that didn´t bother him any. He was used to it, and there is no doubt he liked the power he had of making his fellow creatures fly around. When he found, sitting on a branch, with two toes front and two toes back, a downy puff with big round eyes and a Roman nose and feather–horns sticking up like the ears of a cat, maybe he was a bit puzzled because it didn´t fly, too. Perhaps he didn´t quite know what to make of poor little Solomon, who, disturbed from his nap, just drew himself up slim and tall, and remarked, "Oh!" in a sweet high voice.

我们嬉笑玩耍时产生的那种叫快乐的感觉，好像他也有。对于一只一到白天就百无聊赖、昏昏欲睡的鸟儿来说，这也许真的很好玩。总之，他一过去，别的鸟儿就会四散逃窜。即使他知道这是为什么也无所谓，他已经习惯了，而且，毫无疑问，他很高兴自己能有本事让同类们惊慌失措、四散逃窜。可是，他发现有一只鸟竟然没逃走。那个家伙像个小绒球似的蹲在那里，四只爪尖两前两后紧紧地抓着树枝，睁着大大的圆眼睛，挺着高高的直鼻梁，脑袋上的羽毛像猫耳朵似的向上竖着。这下，他可有点儿疑惑了。也许他真的有点儿不知所措了！可怜的小所罗门，刚才还睡得好好的呢。不过，他只绷紧了身子、向上竖着，说了声“噢！”声音很大，但是还算温和。

But, puzzled or not, Jay knew very well what he could do about it. He had done it so many times before! It was a game he liked. He stood on a branch, and called Solomon names in loud, harsh tones. He flew around as if in a terrible temper, screaming at the top of his voice. When he began, there was not another day bird in sight. Before many minutes, all the chickadees, nuthatches, and woodpeckers within hearing had arrived, and had taken sides with Jay. Yes, even sunny-hearted Chick D.D. himself said things to Solomon that were almost saucy. I never heard that any of these mobs actually hurt our little friend; but they certainly disturbed his nap, and there was no peace for him until he slipped away. Where he went, there was no sound to tell, for his feathers were fringed with silent down. Perhaps some snow-bowed branch of evergreen gave him shelter, in a nook where he could see better than the day-eyed birds who tried to follow and then lost track of him.

不过，不管疑惑还是不疑惑，讨厌鬼松鸦都知道怎么施展他的恶作剧。这对他已经是轻车熟路了。这个把戏他喜欢。他落到一根树枝上，扯着嗓子骂起了所罗门,一边骂,一边还绕着所罗门飞来飞去,一副声嘶力竭、气急败坏的样子。他刚开骂的时候,周围一只鸟儿也没有。可没过几分钟,所有的山雀、五子雀和啄木鸟听到他的叫骂声都赶过来为他助威了。是的,连阳光小鸟奇客嘀嘀都冲着所罗门嚷了一通,就差说脏话了。虽然,我从没听说过这些聚众闹事的鸟儿中有谁真的伤害过我们的小朋友所罗门,可是,他们的的确确吵得他无法安睡,所以,他只有溜之大吉,否则无法得到安宁。那么他溜到哪儿去了呢？听是听不出来的,因为他的羽毛上有一圈能静音的绒毛。也许他躲到常青树上被积雪压弯了的树枝上去了。在那里，他比那些鸟儿看得清楚，因为他们的眼睛白天才能看得见。他们想追他,可很快就被甩掉了。

So Solomon went on with his nap, and Jay started off in quest of other adventures. The winter air put a keen edge on his appetite, which was probably the reason why he began to hunt for some of the cupboards where food was stored. Of course, he had tucked a goodly supply of acorns and such things away for himself; but he slipped into one hollow in a tree that was well stocked with frozen fish, which he had certainly had no hand in catching. But what did it matter to the blue-jacketed robber if that fish had meant a three-night fishing at an air-hole in the ice? He didn´t care (and probably didn´t know) who caught it. It tasted good on a frosty day, so he feasted on fish in Solomon´s pantry, while the little owl slept.

于是,所罗门接着睡他的觉,而讨厌鬼松鸦则又去寻别的开心了。冬天的凛冽让他食欲大增,也许就是这个原因,他开始搜寻藏有食物的壁橱。当然了,他已经为自己储备了大量的橡子之类的食物。不过,他还是偷偷溜进了一个空树洞里,那里存着很多冻鱼,不过这些鱼一条也不是他捕的。就算一条鱼就意味着要在冰面的通气孔边连续守候三个晚上,可对这个穿着蓝色外套的强盗来说那又有什么关系呢?他才不在乎(也许是不知道)鱼是谁捉的呢!天寒地冻,鱼的味道不错。所以,他趁着小猫头鹰所罗门呼呼大睡的时候,在他的食物储藏室里美美地吃了一顿。

Well, if Jay, the bold dashing fellow, held noisy revel during the dazzling winter days, night came every once in so often; and then a quavering call, tremulous yet unafraid, told the listening world that an elf of the moonlight was claiming his own. And if some shivered at the sound, others there were who welcomed it as a challenge to enter the realm of a winter´s night.

唉,即使讨厌鬼松鸦这个肆意妄为、惹是生非的坏家伙,整天在明晃晃的冬日里聚众喧闹,可是,夜晚还是会如期降临。到那时,一声尖叫划破夜空,这声音在震颤,却听不出丝毫的畏惧,他在向侧耳聆听的世界宣告:夜的精灵又在索取属于自己的那一份东西。假如有些人听到这个声音浑身战栗的话,那么还有另外一些人听到这个声音则欢欣鼓舞,在他们看来这是进入冬夜王国必须接受的挑战。

For, summer or winter, the night holds much of mystery, close to the heart of which lives a little downy owl, who wings his way silent as a shadow, whither he will. And when he calls, people who love the stars and the wonders they shine down upon sometimes go out to the woods and talk with him, for the words he speaks are not hard even for a human voice to say. There was once a boy, so a great poet tells us, who stood many a time at evening beneath the trees or by the glimmering lake, and called the owls that they might answer him. While he listened, who knows what the bird of wisdom told him about the night?

因为，无论是酷暑还是寒冬，夜都是那么的神秘莫测。在那片神秘莫测的世界里，生活着一只小小的、毛茸茸的猫头鹰，他像影子一样悄无声息地挥动翅膀，恣意飞行。当他鸣叫的时候，那些热爱天上的群星、热爱群星照耀下的奇迹的人们，有时候会走出家门来到树林，和他畅谈。他说的话并不难，人类的声音就能模仿。一位了不起的诗人说，曾经有一个男孩，许许多多个傍晚，他都站在树下，或者伫立在波光粼粼的湖边，呼唤着猫头鹰，而猫头鹰们也会回应他的呼唤。在他侧耳聆听他们的回答时，谁知道那智慧之鸟为他讲述了些什么夜的故事呢？

12

Bob the Vagabond

食米鸟鲍勃，快乐的流浪汉

Bob had on his traveling suit, for a vagabond must go a-journeying. It would never do to stay too long in one place, and here it was August already. Why, he had been in Maine two months and more, and it is small wonder he was getting restless. Restless, though not unhappy! Bob was never that; for the joy of the open way was always before him, and whenever the impulse came, he could set sail and be off.

鲍勃已经换上了旅行装，因为流浪汉就得旅行。在一个地方待太久可绝对不行，再说现在已经八月了。要知道，他已经在缅因州待了两个多月了，也难怪他开始有点焦躁不安了。就是焦躁不安，不是不开心！鲍勃可从来没有不开心过，因为道路广阔无边，随时都在等着他去享受那份旅行的快乐；只要兴致所致，他都可以随时展翅出发。

The meadows of Maine had been his choice for his honeymoon, and a glad time of it he and May had had with their snug little home of woven grass. That home was like an anchor to them both, and held their hearts fast during the days it had taken to make five grown-sized birds out of five eggs. But now that their sons and daughters were strong of wing and fully dressed in traveling suits like their mother´s, it was well that Bob had put off his gay wedding clothes and donned a garb of about the same sort as that worn by the rest of his family; for dull colors are much the best for trips.

他来缅因州，是为了在这里的草地上度蜜月。他和梅在这里织了一个温暖舒适的小草窝，度过了那段快乐的时光。在那段日子里，五只蛋变成了五只大鸟，那个小草窝就像是一根锚，紧紧地拴着他俩的心。不过，既然他们的儿女翅膀已经硬了，而且都全副武装穿上了妈妈那样的旅行服，鲍勃也就脱下了鲜艳的结婚礼服，换上了和家里其他人一样的衣服，因为，暗淡的颜色最适合旅行。

Now that they were properly dressed, there was nothing left to see to, except to join the Band of Bobolink Vagabonds. Of course no one can be a member of this band without the password; but there was nothing about that to worry Bob. When any of them came near, he called, "Chink," and the gathering flock would sing out a cheery "Chink" in reply: and that is the way he and his family were initiated into the Band of Bobolink Vagabonds. Anyone who can say "Chink" may join this merry company. That is, anyone who can pronounce it with just exactly the right sound!

穿戴妥当之后，一切准备就绪，就差加入“食米鸟流浪汉帮”了。当然了，没有口令谁都成不了这个帮的成员；不过对此鲍勃却一点也不担心。只要这个帮里的鸟飞过来，鲍勃都会叫一声“卿客”，聚集而来的鸟群也会欢快地大声回答“卿客”。鲍勃一家就这样加入了“食米鸟流浪汉帮”。无论是谁，只要会说“卿客”就可以融入这个快乐的行列中去。也就是说，无论是谁，只要能准确无误地念出这个词就行！

So, with a flutter of pleasant excitement, they were gone. Off, they were, for a land that lies south of the Amazon, and with no more to say about it than, "Chink".

于是，他们挥动翅膀，欢欣鼓舞地出发了。他们只说了一句“卿客”，就飞呀飞，向亚马孙河南面的一块大地飞去。

No trunk, no ticket, no lunch–box; and the land they would seek was four thousand miles or more away! Poor little Bob! Had he but tapped at the door of Man with his farewell "Chink", someone could have let him see a map of his journey. For men have printed time–tables of the Bobolink Route, with maps to show what way it lies, and with the different Stations marked where food and rest can be found. The names of some of the most important Stations that a bobolink, starting from Maine, should stop at on the way to Brazil and Paraguay, are Maryland, South Carolina, Florida, Cuba, Jamaica, and Venezuela.

没有行李箱，没有车票，也不用午餐盒；他们寻找的那块大地远在四千英里之外，或者更远！可怜的小鲍勃！当初他只要敲敲人类的门，说一声“卿客”，道一声别，那么一定会有人能让他看看他的旅行图。因为人类已经印制了“食米鸟迁徙路线”表，表上有地图，标明了食米鸟的迁徙路线，还有不同的站点，哪里有食物，哪里能休息都标得清清楚楚。如果一只食米鸟从缅因州出发，飞往巴西和巴拉圭，那么沿途他要在下面这些最重要的站点停留：马里兰、南卡罗来纳州、佛罗里达州、古巴、牙买加和委内瑞拉。

Does it seem a pity that the little ignorant bird started off without knowing even the name of one of these places? Ah, no! A journeying bobolink needs no advice. "Poor," indeed! Why, Bob had a gift that made him fortunate beyond the understanding of men. Nature has dealt generously with Man, to be sure, giving him power to build ships for the sea and the air, and trains for the land, whereon he may go, and power to print time-tables to guide the time of travel. But to Bob also, who could do none of these things, Nature had, nevertheless, been generous, and had given him power to go four thousand miles without losing his way, though he had neither chart nor compass.

这个无知的小鸟，他连这些地方的名字一个都不知道就出发了，是不是很可怜呀！噢，不会的！旅行中的食米鸟不需要别人的建议。的确“可怜”！可是，鲍勃有一种天赋，这种天赋让他幸运得连我们人类都无法理解。诚然，大自然对人类很慷慨，赋予了人类建造轮船、飞艇和火车的能力，让他们可以乘坐这些工具穿行在海洋、天空和陆地之间；它还赋予人类印制时间表的能力，让他们借助这些时间表来安排旅行的时间。然而，对鲍勃来说，虽然他不会做这些事情，但是，大自然对他却同样慷慨，它赋予了他飞越四千英里而不会迷路的本领，而且他没有航海图，也没有指南针。

What it would be like to have this gift, we can hardly even guess—we who get lost in the woods a mile from home, and wander in bewildered circles, not knowing where to turn! We can no more know how Bob found his way than the born-deaf can know the sound of a merry tune, or the born-blind can know the look of a sunset sky. Some people think that, besides the five senses given to a man, Nature gave one more to the bobolink—a sixth gift, called a "sense of direction".

有这种天赋会是什么样呢？我们连猜都猜不出来，因为我们在离家一英里之外的树林里就会迷路，只能在里面迷迷糊糊地瞎转，简直就不知道往哪里拐！正如一个生来就耳聋的人无法欣赏欢快的乐曲，或者一个生来就眼瞎的人无法领略日落时的天空一样，我们也无法知道鲍勃是如何找到他的路的。有些人认为，食米鸟除了被赋予了人类的五种感觉能力之外，大自然还多给了食米鸟一种感觉能力——第六种能力，叫作“方向感”。

A wonderful gift for a vagabond! To journey hither and yon with never a fear of being lost! To go forty hundred miles and never miss the way! To sail over land and over sea,—over meadow and forest and mountain,—and reach the homeland, far south of the Amazon, at just the right time! To travel by starlight as well as by sunshine, without once mistaking the path!

对于流浪者来说，这真是一个奇妙的礼物！有了他，无论走到哪里都不会担心迷路！飞越四千英里从不偏离航线！越过陆地、越过海洋——越过草地、森林和山峦——准时抵达远在亚马孙河之南的家乡！无论是沐浴着阳光飞行，还是披着星光赶路，千里迢迢，从没走过一次错路！

By starlight? What, Bob, who had frolicked and chuckled through the bright June days, and dozed o′ nights so quietly that never a passing owl could see a motion to tempt a chase?

什么？披着星光？明媚的六月里，那个整天嬉戏玩耍、轻声欢笑的鲍勃？那个一到晚上就打起瞌睡，安静得连路过的猫头鹰都看不见一点动静、无法追逐的鲍勃？

Yes, when he joined the Band of Bobolink Vagabonds, the gates of the night, which had been closed to him by Sleep, were somehow thrown open, and Bob was free to journey, not only where he would, but when he would—neither darkness nor daylight having power to stop him then.

是的，就是他。加入“食米鸟流浪汉帮”后，被“睡神”关上的“黑夜之门”现在不知为何打开了，此时的鲍勃可以自由自在地旅行了，他不仅想去哪里就去哪里，而且想什么时候去就什么时候去——无论是夜色还是日光都无法阻止他。

Is it strange that his wings quivered with the joy of voyaging as surely as the sails of a boat tighten in the tugging winds?

是不是很奇怪？扬起的船帆一定会在劲风中抖动，鲍勃的翅膀也因为远行的兴奋而颤抖不已。

What would you give to see this miracle—a bobolink flying through the night? For it has been seen; there being men who go and watch, when their calendars tell them it is time for birds to take their southward flight. Their eyes are too feeble to see such sights unaided; so they look through a telescope toward the full round moon, and then they can see the birds that pass between them and the light. Like a procession they go—the bobolinks and other migrants, too; for the night sky is filled with travelers when birds fly south.

看到这个奇迹——一只食米鸟在茫茫夜色中飞翔——你会有什么感想呢？因为有人已经看到了这个奇迹。当日历翻到鸟儿开始南飞的那页时，他们就走出家门，仰天观望。他们的视力太弱，不借助仪器就无法看到这样的景观，所以，他们就透过一架望远镜向那轮圆圆的满月望过去，然后，他们才能看到在他们和月光之间穿越而过的鸟儿。他们像一只前行的队伍，有食米鸟，也有其他候鸟；因为，每当鸟儿南飞的时候，整个夜空都挤满了行色匆匆的旅行者。

But though we could not see them, we should know when they are on their way because of their voices. What would you give to hear this miracle—a bobolink calling his watchword through the night? For it has been heard; there being men who go to the hilltops and listen.

尽管我们无法用肉眼看见他们，可是，因为他们的叫声，我们也能知道他们已经上路了。听到这个奇迹——一只食米鸟在茫茫的夜色中喊着口令——你会有什么感想呢？因为已经有人爬上小山顶，侧耳聆听，听到了这个奇迹。

As they hear, now and again, wanderers far above them calling, "Chink," one to another, they know the bobolinks are on their way to a land that lies south of the Amazon, and that neither sleep nor darkness bars their path, which is open before them to take when and where they will.

当他们时不时地听到这些四处漂泊的鸟儿们在远远的上空互相叫着“卿客”的时候，他们就知道食米鸟已经上路了，他们要飞往亚马孙河以南的那片土地，而且是星夜兼程，因为他们面前的道路畅通无阻，随时随处都可以上路。

And yet Bob and his comrades did not hasten. The year was long enough for pleasure by the way. He and May had worked busily to bring up a family of five fine sons and daughters early in the summer; and now that their children were able to look out for themselves, there was no reason why the birds should not have some idle, care-free hours.

不过，鲍勃和伙伴们并不着急，再说有一年的时间呢，足够好好玩乐一番的了。整个初夏，鲍勃和梅都在忙着哺育家里五个可爱的儿女，现在既然孩子们能自己照顾自己了，他们也就可以享受一些懒散、无忧的时光了。

Besides, it was time for the Feast of the Vagabonds,

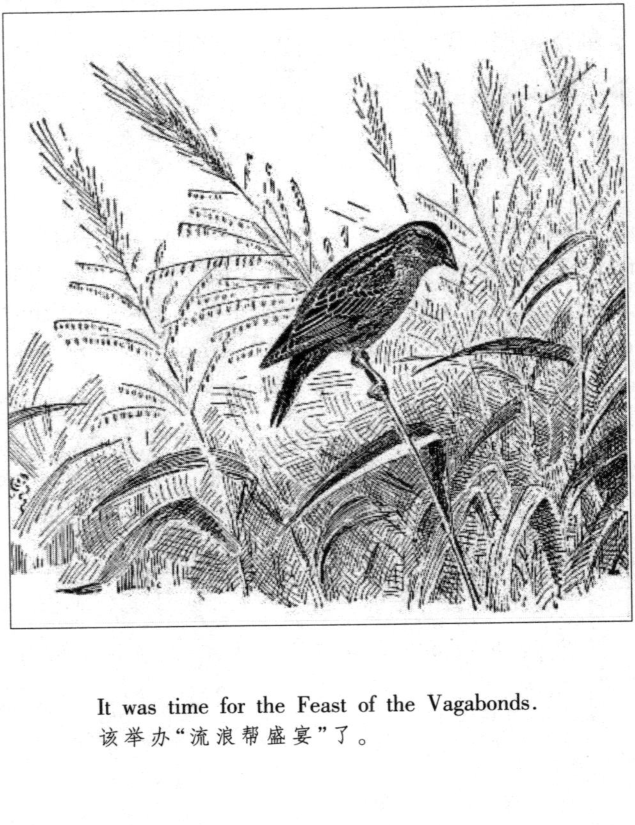

It was time for the Feast of the Vagabonds.
该举办“流浪帮盛宴”了。

a ceremony that must be performed during the first weeks of the Migrant Flight; for it is a custom of the bobolinks, come down to them through no one knows how many centuries, to hold a farewell feast before leaving North America. If you will glance at a map of the Bobolink Route, you will see the names of the states they passed through. Our travelers did not know these names; but for all that, they found the Great Rice Trail and followed it. They found wild rice in the swamps of Maryland and the neighboring states. In South Carolina they found acres of cultivated rice. For rice is the favorite food during the Feast of the Vagabonds, and to them Nature has a special way of serving it. This same grain is eaten in many lands; taken in one way or another, it is said to be the principal food of about one half of all the people in the world. Bob didn´t eat his in soup or pudding or chop–suey. He used neither spoon nor chop–sticks. He took his in the good old–fashioned way of his own folk—unripe, as most of us take our sweet corn, green and in the tender, milky stage, fresh from the stalk. He had been having a rather heavy meat diet in Maine, the meadow insects being abundant, and he relished the change. There was doubtless a good healthy reason for the ceremony of the Feast of the Vagabonds, as anyone who saw Bob may have guessed; for by the time he left

South Carolina he was as fat as butter.

另外，该举办“流浪帮盛宴”了，这个仪式必须在“迁徙之旅”开始后的前几周举行。离开北美洲之前举办告别宴会是食米鸟流传了很多个世纪的习惯，到底有多少个世纪谁也不知道。如果你看一眼“食米鸟迁徙路线”图，就会看到他们经过的那些州的州名。可是，除此之外，他们还发现了“伟大的稻米路”，并且沿着它一路前行。他们在马里兰州和附近的几个州的沼泽地里发现了野生水稻；在南卡罗莱纳州发现了一亩亩的栽培水稻。稻米是“流浪帮宴会”上最受欢迎的食物，而且大自然自有其特别的方式把稻米送到他们面前。很多地方都吃这种粮食，据说，世界上大约有一半的人以它为主食，不过吃的方法各式各样。鲍勃吃稻米的时候，不用做成粥、米糕或者炒饭，也用不着勺子或者筷子。他吃稻米的方法和其他食米鸟的方法一样，已经过时，但却有益健康：趁嫩生吃。就像我们大多数人吃甜玉米那样，趁着乳熟期的嫩绿，从秆上摘下来直接吃。在缅因州的时候，草地里的昆虫极其丰富，鲍勃吃的一直都是荤食，所以，他很想变换一下口味。毫无疑问，举办“流浪帮宴会”还有健康方面的原因，这个只要见到鲍勃就猜得出来，因为鲍勃离开南卡罗莱纳州时，已经肥得像黄油一样了。

In following the Great Rice Trail, Bob went over the same road that he had taken the spring before when he was northward bound; but one could hardly believe him to be the same bird, for he looked different and he acted differently. In the late summer, the departing bird was dull of hue and, except for a few notes that once in a great while escaped him, like some nearly forgotten echo of the spring, he had no more music in him than his mate, May. And when they went southward, they went all together—the fathers and mothers and sons and daughters in one great company.

沿着“伟大的稻米路”南迁时，鲍勃路过的地方和春天北归时路过的地方一样，可是，人们很难相信那只鸟就是鲍勃，因为他不仅模样变了，举止也不同了。夏末出发的时候，他还是灰秃秃的，而且，和他的伴侣梅一样，身上的音乐细胞也没有了，偶尔鸣叫几声，也是千载难逢的事情，就像是山谷里回荡的报春曲，几乎被淡忘了。南迁时，他们举家同行——爸爸、妈妈、儿子、女儿，浩浩荡荡。

In the spring it had all been different: Bob had come north with his vagabond brothers a bit ahead of the sister-folk. And the vagabond brothers had been gay of garb—fresh black and white, with a touch of buff. And Bob and his band had been gay of voice. The flock of them had gathered in tree-tops and flooded the day with such mellow, laughing melodies as the world can have only in springtime—and only as long as the bobolinks last.

可是到了春天，情形就迥然不同了：鲍勃和流浪帮的兄弟们比姊妹们先行一步踏上北归之路。而且，流浪帮兄弟们的衣着光鲜亮丽——崭新的羽毛黑白相间，还夹杂着一丝鹅黄。他们的鸣叫也清脆响亮。他们整天成群结队地聚集在树梢引吭歌唱，歌声婉转悠扬、轻松明快，响彻天空，这是春天独有的歌声，只要食米鸟不消失，这歌声也不会消失。

The ways of the springtime are for the spring, and those of the autumn for the fall of the year. So Bob, who, when northward bound a few months before, had taken part in the grand Festival of Song, now that he was southward bound, partook of the great Feast of the Vagabonds, giving himself whole-heartedly to each ceremony in turn, as a bobolink should, for such are the time-honored customs of his folk.

春天有春天的事情，秋天有秋天的任务，食米鸟做事泾渭分明。所以，几个月前北归时，鲍勃在盛大的“音乐节”里引吭高歌；如今要南迁了，他就尽情享受“流浪帮盛宴”上的美味佳肴。每次活动他都尽心尽力、全力以赴。作为一只食米鸟就该这样，这可是他们的优良传统。

Honored for how long a time we do not know. Longer than the memory of man has known the rice-fields of South Carolina! Days long before that, when elephants trod upon that ground, did those great beasts hear the spring song of the bobolinks? Is the answer to that question buried in the rocks with the elephants? Bob didn´t know. He flew over, with never a thought in his little head but for the Great Rice Trail leading him southward to Florida.

这个优良传统的历史到底有多悠久，我们不得而知。肯定比人类发现南卡罗莱纳州的稻田要悠久！那么，在比那更久远的时代，当大象迈着沉重的脚步从这片土地走过时，他们是否听到过食米鸟的春之歌？这个问题的答案是否也和大象一样被埋在了岩石下面？鲍勃不知道。他飞过去的时候，小脑袋瓜里只有一个念头：沿着“伟大的稻米路”往南飞，一直飞到佛罗里达。

While there, some travelers would have gone about and watched men cut sponges, and have found out why Florida has a Spanish name. But not Bob! The Feast of the Vagabonds, which had lasted well-nigh all the way from Maryland, was still being observed, and even the stupidest person can see that rice is better to eat than sponges or history.

有些旅行者在佛罗里达滞留期间，会四处游走，看到人类采割海绵，于是就会发现佛罗里达为什么会有一个西班牙名字。可是，鲍勃不会！从马里兰一路飞来，他们整天都在举办“流浪帮盛宴”，到现在还没结束，连最笨的人都知道稻米比海绵好吃得多，也比研究历史实惠得多。

Then, as suddenly as if their "Chink, chink, chink" meant "One, two, three, away we go," the long feast was over, and their great flight again called them to wing their way into the night. How they found Cuba through the darkness, without knowing one star from another; what brought them to an island in the midst of the water that was everywhere alike—no man knows. But in Cuba they landed in good health and spirits. This was in September,—a very satisfactory time for a bird–visit,—and Bob and his comrades spent some little time there, it being October, indeed, when they arrived on the island of Jamaica. Now Jamaica, so people say who know the place, has a comfortable climate and thrilling views; but it didn´t satisfy Bob. Not for long! Something south of the Amazon kept calling to him. Something that had called to his father and to his grandfather and to all his ancestors, ever since bobolinks first flew from North America to South America once every year.

突然之间,这场旷日持久的盛宴结束了,他们"卿客,卿客,卿客"的叫声突然之间也好像一下子变成了"一二三,出发"的口令。伟大的迁徙之旅又在呼唤,他们张开双翅,飞进了茫茫的夜色里。他们对星星一无所知,他们是如何越过漫漫黑夜找到古巴的呢?岛屿四周海水茫茫、处处相似,是什么引导他们到达的呢?无人知晓。可是抵达古巴的时候,他们个个身强体壮、精神抖擞。时至九月,正是鸟儿们造访的佳期,鲍勃和他的伙伴们在那里小住了一阵,所以到达牙买加岛的时候,已经十月了。知道牙买加的人都说那里的气候宜人、景色摄人心魄,可是,即使这样也留不住鲍勃的心,他待不了多久!亚马孙河以南有种东西在不断地呼唤着它;一年一度,当食米鸟先从北美洲出发飞向南美洲的时候，这种东西就在不断地呼唤着他；他也曾呼唤过他的祖祖辈辈，包括他的爷爷和他的爸爸。

How many ages this has been, who knows? Perhaps ever since the icy glaciers left Maine and made a chance for summer meadows there. Long, long, long, it has been, that something south of the Amazon has called to bobolinks and brought them on their way in the fall of the year. So the same impulse quickened Bob´s heart that had stirred all his fathers, back through countless seasons. The same quiver for flight came to all the Band of Vagabonds. Was it homesickness? We do not know.

到底呼唤了多久，谁知道呢？也许早在刺骨的冰川离开缅因州，夏季的草场乘虚而入的时候就开始了。很久、很久以来，亚马孙河以南有个东西就一直在呼唤食米鸟，让他们一入秋就踏上漫漫征程。春去秋来，季节变换无数，早在无数年之前，鲍勃的父辈们就因出发的冲动而兴奋不已，现在鲍勃也按捺不住这种冲动，心越跳越快。这种渴望飞翔的冲动席卷了整个流浪帮，让他们浑身战栗。他们是得了思乡病吗？我们不知道。

Something south of the Amazon kept calling to him.
亚马孙以南有一种东西在不断地呼唤着他。

We only know that a night came when Bob and his companions left the mountains of Jamaica below them and then behind them. Far, far behind them lay the island, and far, far ahead the coast they sought. Five hundred miles between Jamaica and a chance for rest or food. Five hundred miles; and the night lay about and above them and the waters lay underneath. The stars shone clear, but they knew not one from another.

我们只知道，当鲍勃和他的伙伴们越过牙买加的崇山峻岭，把他们远远抛在身后时，夜幕就降临了。他们身后很远、很远的地方是那个海岛，前方很远、很远的地方是他们朝思暮想的海岸。离开牙买加后，要再飞五百英里才有机会休息、觅食。五百英里的路途，头顶和四周是茫茫黑夜，身下是无边的海水。群星闪烁，可是他们却一颗也不认识。

No guide, no pilot, no compass, such as we can understand, gave aid through the hours of their flight. But do you think they were afraid? Afraid of the dark, of the water, of the miles? Listen, in your fancy, and hear them call to one another. "Chink," they say; and though we do not know just what this means, we can tell from the sound that it is not a note of fear. And why fear? There was no storm to buffet them that night. They passed near no dazzling lighthouse, to bewilder them. No danger threatened, and something called them straight and steady on their way.

没有向导,没有引航员,没有指南针,在他们的漫漫飞行之旅中,没有我们熟知的这一切为他们指引方向。你以为他们会害怕吗?害怕黑暗,害怕海水,害怕那漫长的飞行之路?在你神思飞扬的时候,听听他们的鸣叫吧!你会听到他们在相互呼唤,不住地说着"卿客",虽然我们不知道这是什么意思,但是听声音,我们就知道这不是懦弱的哀鸣。为什么要害怕呢?这个夜晚风平浪静,没有暴风雨的肆虐;所经之处,也没有灯塔炫目的光柱让他们头昏目眩;没有危险,没有威胁,只有那让他们勇往直前的呼唤。

Oh, they were wonderful, that band! Perhaps among all living creatures of the world there is nothing more wonderful than a bird in his migrant flight—a bird whose blood is fresh with the air he breathes as only a bird can breathe; whose health is strong with the wholesome feast that he takes when and where he finds it; whose wings hold him in perfect flight through unweary miles; whose life is led, we know not how, on, on, on, and ever in the right direction.

哦，太棒了，这群食米鸟！也许，世界上所有的生命中，只有迁徙途中的鸟儿最棒！鸟儿独特的呼吸方式，让他浑身热血奔涌、充满了活力；随时随地只要有食就大吃特吃的习惯让他体质强健；那强劲有力的翅膀让他越过了千山万水却不知疲倦；我们不知道，究竟是什么力量引导着他的生命之旅，让他一直沿着正确的方向向前，向前，再向前。

Yes, Bob was wonderful when he flew from the mountains of Jamaica to the great savannas of Venezuela; but he made no fuss about it—seemed to feel no special pride. All he said was, "Chink," in the same matter-of-fact way that his bobolink forefathers had spoken, back through all the years when they, too, had taken this same flight over sea in the course of their vagabond journey.

是的，鲍勃很棒！他越过了牙买加的崇山峻岭，来到了委内瑞拉的稀树大草原。不过，他并没有因此而四处炫耀——他似乎没觉得有什么可骄傲的，只是轻描淡写地说了一声“卿客”。很多很多年以来，当他的食米鸟父辈们在他们的漂泊之旅中越过这片大海的时候，也是这样一副轻描淡写的样子。

From Venezuela to Paraguay there was no more ocean to cross, and there were frequent places for rest when Bob and his band desired. Groves there were, strange groves—some where Brazil nuts grew, and some where oranges were as common as apples in New England. There were chocolate trees and banana palms. There were pepper bushes, gay as our holly trees at Christmastime. Great flowering trees held out their blossom cups to brilliant hummingbirds hovering by hundreds all about them. Was there one among them with a ruby throat, like that of the hummingbird who feasted in the Cardinal-Flower Path near Peter Piper´s home? Maybe it was the self-same bird—who knows? And let´s see—Peter Piper himself would be coming soon, would he not, to teeter and picnic along some pleasant Brazilian shore?

从委内瑞拉到巴拉圭，沿途不需要再穿越任何海洋，鲍勃和他的队伍随时可以找到地方就地休息。那里有很多果树林，非常奇特的果树林——有的结着巴西坚果，有的挂满了橘子，到处都是，普通的就像是新英格兰的苹果一样。还有巧克力树和香蕉树，还有胡椒丛，鲜艳夺目，就像是圣诞时我们准备的冬青树一般。只开花不结果的树高大粗壮，怒放的花朵像举起的杯子，让数百只色彩艳丽的蜂鸟流连忘返。其中有一只的脖颈红红的，像宝石一样，是不是很像那只在彼得派普家附近的“红衣半边莲小径”上美餐的蜂鸟？也许就是那只吧，谁知道呢？让我们瞧瞧吧，说不定彼得派普自己一会儿也会飞过来呢。巴西海岸如此令人陶醉，难道它不会追寻而至，沿着海岸摇摇摆摆、觅食野餐吗？

Perhaps Bob and Peter and the hummingbird, who had been summer neighbors in North America, would meet again now and then in that far south country. But I do not think they would know each other if they did. They had all seemed too busy with their own affairs to get acquainted.

夏天在北美的时候，鲍勃、彼得，还有那只蜂鸟曾经是邻居，也许在遥远的南方他们偶尔也会不期而遇吧？不过，我想他们即使相遇也不会认识对方。一直以来，他们似乎都在忙自己的事情，太忙了，根本没时间去结识邻居。

Besides the groves where the nuts and fruit and flowers grew, the vagabonds passed over forests so dense and tangled that Bob caught never a glimpse of the monkeys playing there: big brown ones, with heads of hair that looked like wigs, and tiny white ones, timid and gentle, and other kinds, too, all of them being very wise in their wild ways—as wise, perhaps, as a hand-organ monkey, and much, much happier.

除了长满坚果、水果和开满花儿的那些小树林，这些流浪者还经过了一些大森林，那里的树木密密麻麻、纵横交错，猴子们在里面尽情嬉戏玩耍，可是鲍勃连他们的影子都没看到。那些猴子，有的是棕色的大猴子，头上好像戴着假发套；有些是白色的小猴子，怯生生的很温顺；当然还有其他各种不同的猴子，别看他们那么顽皮，其实他们都聪明极了，也许和会摇手风琴的猴子一样聪明，而且比他们快乐多了。

No, I don´t think Bob saw the monkeys, but he must have caught glimpses of some members of the Parrot Family, for there were so many of them; and I´m sure he heard the racket they made when they talked together. One kind had feathers soft as the blue of a pale hyacinth flower, and a beak strong enough to crush nuts so hard–shelled that a man could not easily crack them with a hammer. But all that was as nothing to Bob, for it was not grove or forest or beast or bird that the vagabonds were seeking.

是呀，我觉得鲍勃没看见那些猴子，不过他一定瞥见过鹦鹉家族的有些成员，因为他们实在太多啦！而且我敢肯定，鲍勃还听到了他们在一起的时候叽叽喳喳的说话声呢。其中有一种鹦鹉，羽毛软软的，像风信子那淡淡的蓝色花瓣；他们的嘴巴太厉害了，人类用斧子都无法轻易砸碎的坚果，他们都能啄个粉碎。可是，所有这一切对鲍勃来说都不算什么，他们这些流浪者日思夜想的不是树丛，不是森林，也不是兽儿和鸟儿。

When they had crossed the Amazon River, some of the band stopped in places that seemed inviting. But Bob and the rest of the company went on till they crossed the Paraguay River; and there, in the western part of that country, they made themselves at home. A strange, topsy–turvy land it is—as queer in some ways as the Wonderland Alice entered when she went through the Looking–Glass; for in Paraguay January comes in the middle of summer; and the hot, muggy winds blow from the north; and the cool, refreshing breezes come from the south; and some of the wood is so heavy that it will not float in water; and the people make tea with dried holly leaves! But to the Band of Vagabond Bobolinks it was not topsy-turvy, for it was home; and they found the Paraguay prairies as well suited to the comforts of their January summer as the meadows of the North had been for their summer of June.

越过亚马孙河后,有些食米鸟就在看上去比较诱人的地方停了下来。不过鲍勃和其他伙伴们继续向前飞去,一直飞过了巴拉圭河,来到了巴拉圭国的西部,他们在那里安营扎寨,舒舒服服地过起了日子。这是一片奇异的国土,一切都颠三倒四的,没有章法,有些方面奇怪得就像当初爱丽丝穿过镜子偶然闯入的那个奇境一样。巴拉圭的一月正值仲夏,北边吹来的是又闷又热的大风,南边飘来的是清新凉爽的微风;有些木头沉得放到水里都浮不起来;这里的人竟然把冬青树叶晒干泡茶喝!但是对于食米鸟流浪帮来说,这很正常,因为这里是家。他们觉得,巴拉圭一月夏季的大草原和北方六月夏季的草地一样,都很舒服。

Bob was satisfied. He had flown four thousand

miles from a meadow and had found a prairie! And if, in all that wonderful journey, he had not paid over much attention to anything along the way except swamps and marshes, do not scorn him for that. Remember always that Bob *found* his prairie and that Peter *found* his shore.

鲍勃心满意足！他离开一块小草地，一路飞行四千英里，结果发现了一个大草原！如果说，在这次无与伦比的旅行中，除了沼泽和湿地之外，沿途他什么都不曾留意，请不要笑话他。请你永远铭记这个事实：鲍勃"找到了"他的大草原，彼得派普"找到了"他的岸。

It is somewhere written, "Seek and ye shall find." It is so with the children of birds—they find what Nature has given them to seek. And is it so with the children of men? Never think that Nature has been less kind to boys and girls than to birds. Unto Bob was given the fields to seek, and he had no other choice. Unto Peter the shores, and that was all. But unto us is given a chance to choose what we will seek. If it is as far away as the prairies of Paraguay, shall we let a dauntless little vagabond put our faith to shame? If it is as near as our next-door meadow, shall we not find a full measure of happiness there—mixed with the bobolink´s music of June?

曾经有人写过这样一句话："假如你去找，你就会找到。"对于鸟类的孩子来说就是如此。他们找到了大自然妈妈让他们找的东西。那么，对人类的孩子来说也是如此吗？你可千万别以为大自然妈妈只偏爱鸟类，而不垂青人类的男男女女呀。鲍勃奉命去找广阔的田野，他别无选择；彼得奉命去找湖滨和海岸，仅此而已；可是，我们得到的是机会，想找什么就找什么的机会。假如我们要找的地方很远很远，远到就是巴拉圭的大草原，那么我们的信念会不会输给意志坚定的食米鸟呢？假如我们要找的地方很近很近，近到就是隔壁的草坪，那么我们会不会在那里找到极度的幸福，还有食米鸟六月的歌声呢？

For Bob comes back to the North again, bringing with him springtime melodies, which poets sing about but no human voice can mimic. Bob, who has dusted the dull tips from his feathers as he flew, and who, garbed for the brightness of our June, makes a joyful sound; for Nature has kept faith with him and brought him safely back to his meadow, though the journey

Nature has kept faith with him and brought him safely back to his meadow.
大自然妈妈相信他，让他安全地回到生活过的草地。

from and to it numbered eight thousand miles!

因为鲍勃又回到了北方，他带回了春天的旋律，这是让诗人们赞美不已的旋律，是人类无法模仿的旋律。为迎接人类六月的明媚，鲍勃在北归途中就已经一边飞一边掸掉了羽毛尖上的土灰，披上了鲜艳的盛装。他欢快地鸣叫着，因为虽然路途遥远，来回长达八千英里，但是大自然妈妈还是遵守了许下的诺言，将他安全地带回了曾经的草地。

His trail is the open lane of the air,
and the winds, they call him everywhere;
So he wings him North, dear burbling Bob,
with throat aquiver and heart athrob;
And he sings o´ joy in the month of June,
Enough to keep the year in tune.

无边无际的天空就是他的航线，
到处都是风儿的呼唤；
可爱的鲍勃，展翅北飞、嘟嘟囔囔，
带着颤动的喉咙，怦怦跳动的心；
他在歌唱六月的快乐，
快乐就成了一年的基调。

Then, when the rollicking young of his kind
Yearn for the paths that the vagabonds find,
He leads them out over loitering ways
Where the Southland beckons with luring days;
To wait till the laughter–like lilt of his song
is ripe for the North again—missing him long!

突然间欢闹嬉戏的小家伙们，
魂牵梦绕起流浪者们曾经的征程，
他引领着孩子们踏上了漂泊之路，
南国那诱人的日子在那里频频招手；
等着吧，等着他如笑的欢歌
再次回荡在北国的上空——将他久久思念！